General Engineering Chemistry

Dr.D. Gajalakshmi

Published by

General Engineering Chemistry

Copyright © 2017 by Bonfring

ISBN 978-93-86638-13-7

Author

Dr.D. Gajalakshmi

Bonfring

309, 2nd Floor, 5th Street Extension, Gandhipuram,

Coimbatore-641 012.

Tamilnadu, India.

E-mail: info@bonfring.org | Website: www.bonfring.org

Phone: 0422 4213231

Dedicated to my Husband

Dr.T. Sivakumar and my Beloved Parents

Preface

The book on General Engineering Chemistry is written for the benefit of B.E/B.Tech students to enhance their knowledge about the key concepts in Engineering Chemistry. It covers the basic concepts of various Engineering applications. This book enhances the facts in the molar calculations for gaseous reactions involved in thermodynamics with suitable simple examples. The basic idea about Bio Informatics needs to known by understanding the simple carbohydrate structure and their functions in metabolism. This book addresses the structure, concepts of carbohydrate metabolism and stages of glycolysis. Preparations of metal complexes are emphasized with simple examples. This text of General Engineering Chemistry covers not only the Engineering Chemistry aspects, it also covers the fundamental concepts needed in the research areas of materials science. The author welcomes suggestions for further improvement of this book.

Dr.D. Gajalakshmi

Author Profile

Dr.D. Gajalakshmi is now working as Assistant Professor in the Department of Chemistry, University College of Engineering (a constituent College of Anna University Chennai) Villupuram. She obtained her Doctorate from Anna University Chennai and M.Phil Degree from the University of Madras, Guindy Campus. She has 10 years of teaching experience. With 2 years industry exposure, She has got indepth understanding about chemical testing of metals (ferrous and non –ferrous alloys) as per IS, ASTM and BIS standards, Intergranular corrosion testing as per ASTM methods. She got trained in testing Paints, Cements, Fuses, marbles. She is also exposed to procedures and clauses involved in national level accreditation of testing laboratory (**National Accreditation Board for Testing Laboratory (NABL) Delhi**). She has published papers in reputed journals like RSC, Elsevier and Taylor and Francis and presented papers in international conferences. Her area of expertise are Sensor Materials, Plant derived Nano Materials for Sensor, Bio medical applications, Modelling molecules for Optoelectronic Applications (Solar cell and Dye sensitized Solar Cell).

<table>
<tr><th>Chapters</th><th>Contents</th><th>Page No</th></tr>
</table>

Chapters	Contents	Page No
1	**Basics of Chemistry**	**1**
	1. Introduction	1
	1.1. Matter	1
	1.2. Types of Properties of Matter	2
	1.3. Metallic, Semi-metallic, and Non-metallic	4
	1.4. Acids, Bases, and Salts	4
	1.5. Molecular Mass	6
	1.6. Elements	10
	1.7. Electronic Structure of Atoms	10
	1.8. Elements and their Isotopes	13
	1.9. Molecular Formula	17
	1.10. Molecular Mass	18
	1.11. Mole	18
	1.12. Basic Laws and Phenomena of Chemistry	19
	1.13. Properties of Gases	19
	1.14. The Kinetic-Molecular Gas Theory	20
	1.15. Elastic and Inelastic Scattering	20
	1.16. Basic Gas Laws	21
	1.17. Chemical Bonding	23
	1.18. Hydrogen Bonding	24
	1.19. Chemical Reactions	25
	1.20. Oxidation (Redox) Reactions	26
	1.21. The Concept of the Mole	27
	1.22. Stoichiometric Calculations	28
	1.23. Types of Homogeneous Mixtures	28
	1.24. Solution Properties	29
	1.25. Stages of Chemical Equations	31
	1.26. Reduction-Oxidation (Redox) Reactions	32
	1.27. Molarity	32
	1.28. Mole Fractions	34
	1.29. Colligative Properties	35
	Questions	37
2	**Basic Organic Chemistry**	**38**
	2.1. Organic Molecules	38
	2.2. Organic Compound Families	38

2.3. The Alkanes 38

2.4. Drawing Conventions for Organic Molecules 42

2.5. Properties of Alkanes 45

2.6. Combustion of Alkanes 45

2.7. The Alkenes, Alkynes, and Aromatics 46

2.8. Hydrogenation 50

2.9. Aromatic Compounds 50

2.10. Functional Groups: Alcohols, Ethers, Aldehydes, and Ketones 54

2.11. Classification of Alcohols 56

2.12. Functional Groups: Carboxylic Acids, Esters, Amines, and Amides 61

2.13. The Concept of Aromaticity 71

2.14. The Concepts of Saturation and Unsaturation 72

Questions 74

3   Complex Organic Molecules   75

3.1. Carbohydrates: Sugars to Polysaccharides 75

3.2. Carbohydrates: Cellulose and Glycogen 80

3.3. Lipids: Fatty Acids and Waxes 81

3.4. Lipids: Triacylglycerols to Glycerophospholipids 84

3.5. Steroids 88

Questions 92

4   Inorganic Chemistry   93

4.1. Hydrogen and Hydrides 93

4.2. Main group Elements of 2nd and 3rd Periods and their Compounds 97

4.3. Oxygen and Oxides 105

4.4. Chalcogen and Chalcogenides 123

4.5. Halogens and Halides 126

4.6. Rare Gases and their Compounds 135

4.7. Reaction and Physical Properties 137

Questions 148

5   Applications of Chemistry   150

5.1. Thermo Chemistry 150

5.2. Electrochemistry 156

5.3. Nuclear Chemistry 164

5.4. Basic Biological Chemistry 169

Questions 197

CHAPTER 1

BASICS OF CHEMISTRY

1. Introduction

The universe is made up of matter. The matter is anything that occupies space and has mass or weight. The matter is made up of fundamental particles called elements. To further understand chemistry the basic concepts are essential.

1.1. Matter

There are three main states of matter: *solid*, *liquid*, and *gas*. A solid has its own volume and shape. A liquid has its own volume but takes the shape of the vessel it is in. A gas fi the vessel it is in–it takes up the vessel's volume and shape. (A powder takes the shape of the vessel it is in, but the individual grains have their own shape.)

Solid materials diff from one another in size or shape (e.g. steel plate, steel rod, steel wire, and steel wool), and in what they are made of (e.g. steel wire and copper wire). Materials differ only in size or shape are said to be made of the same *basic* material (e.g. "steel", or more precisely, "mild steel", the ordinary kind of steel).

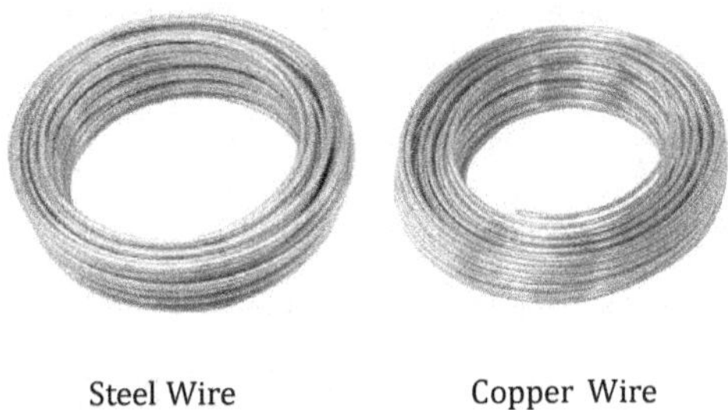

Figure 1: Types of Materials

- Matter is made up of atoms.
- The atoms of an element are chemically all the same, and diff from those of other elements. Atoms are represented by symbols (e.g. O for an oxygen atom and H for a hydrogen atom).
- Chemical reactions involve changes in the way in which atoms are combined, but not in their numbers.
- Atoms of different elements often combine in definite ratios.

- The ratios are often of small whole numbers.
- At low pressures, equal volumes of gases at the same temperature and pressure contain equal numbers of particles (Avogadro's hypothesis). The particles may be atoms or clusters of atoms joined together ("molecules", meaning "little masses", from Latin *moles*, "mass").

Many solids are crystalline, with plane faces. They can be explained by the regular packing of small particles. A familiar example of such packing is in snooker, where the red balls are made to form of a triangle.

Figure 2: Crystalline Formation

Gases are much more compressible than liquids or solids, and when they condense there is a large reduction of volume. The observations can be explained if gases comprise separate particles, which come together in the liquid or solid state.

When a small quantity of olive oil is poured on to water, the oil only spreads over a limited area of the surface.

1.2.   Types of Properties of Matter

Matter can be described as anything that has mass and occupies space.

However, when studying a fundamental science like chemistry it is necessary to make more distinct classifications of matter.

The first major classification of matter is the division of ***physical*** and ***chemical*** properties.

The ***physical properties*** of a substance are those properties that do not depend on a chemical change in the substance in order to be defined.

Some examples of physical properties are listed below.

- Mass
- Volume

- Color
- Freezing Point
- Boiling Point
- Viscosity

The ***chemical properties*** of a substance are those properties that depend on a chemical change or reaction to occur in order to be defined.

Some examples of chemical properties are listed below.

- Corrosion Rate
- Heat of Combustion
- Enthalpy of Formation
- Electromotive Force
- Toxicity
- Reactivity with Solvents

When asked to identify a property of a substance as either physical or chemical simply decide whether a reaction or change is necessary in order to measure the property. If reaction or change is necessary, the property is chemical in nature. If no reaction or change is necessary, the property is physical in nature.

Another way that chemical properties differ from physical properties is that chemical properties can be used to classify a substance. Different materials can be ranked or classified according to a chemical property.

For example, by measuring flammability of various materials, one can classify them as being flammable or inflammable.

Another important distinction of matter is the distinction between ***intensive*** and ***extensive*** properties.

An ***intensive*** property of a substance is a property that does not depend on the amount of the substance present. Some examples of intensive properties would be ***color, odor, density,*** and ***electrical conductivity.*** In other words, a single red brick would be the same color as a ton of the same red brick.

An ***extensive*** property of a substance is a property that does depend on the amount of the substance present. Some examples of extensive properties are ***mass, volume,*** and ***weight.*** If you add one red brick to your system containing one red brick, you double the system's mass, volume, and weight.

A final and more fundamental distinction of a substance is whether a substance is an **element** or whether the substance is a **molecule.**

An **element** is substances that cannot be broken down and retain the identity of the substance. Think of a "pure" gold ring. The ring is comprised of nothing but gold **atoms.** If you were to succeed in breaking apart a gold atom it would cease to be gold.

A **molecule** is a substance comprised of a fixed proportion of two or more atoms. Water is a molecule comprised of two hydrogen atoms and one oxygen atom. If a molecule is comprised of a different ratio of hydrogen and oxygen atoms, it is a different molecule.

Organic and Inorganic

As we have seen, "organic" materials are those derived, directly or indirectly, from plants and animals. "Inorganic" materials are those derived from minerals.

1.3. Metallic, Semi-metallic, and Non-metallic

As mentioned earlier, metals are characterized by being lustrous and very good conductors of electricity. Non-metallic materials are the opposite: they are dull and very poor conductors of electricity. Semi-metallic materials have intermediate properties. The class includes semiconductors.

When this classify is used, the conditions need to be specific Materials can change in character with change of temperature or pressure. Some non-metallic materials (e.g. common salt) conduct electricity in the molten state and in water with the formation of new materials at the electrodes. Materials that do this may be described as "electrolytic".

1.4. Acids, Bases, and Salts

As we have seen, "acids" have a sour taste. They also have the property of turning litmus (a dye extracted from lichens) red.

Acids vary in their ability to dissolve metals. The strongest dissolve many metals easily. They include sulfuric, hydrochloric, and nitric acids. Organic acids are generally weaker.

In their reactions with acids, metal oxides, hydroxides, and carbonates function as a base

$$\text{acid + base salt + secondary products}$$

The secondary products are water and, for carbonates, carbonic oxide. Since a metal carbonate is itself a salt, the salt of a weak acid, the reaction in this case can also be thought of as a displacement:

salt of weak acid + strong acid, salt of strong acid + weak acid

"Bases" are materials that "neutralize" acids, i.e. take away their distinctive properties. Bases that are soluble in water are called "alkalis" (Arabic al-ḳalī, "the ashes"). Examples are soda-ash, caustic soda, potash, and slaked lime. Solutions of alkalis feel soapy, and turn litmus blue.

What chemists call "salts" are the main products that are formed when an acid is neutralized by a base. Chemists these "salts" because they are similar to common salt, which itself can be made by the neutralization of an acid by a base (Chap. 4). Other products are water and, when the base is soda-ash or potash.

1.4.1. Naming Acids and Bases

Acids Without Oxygen in the Anion

If an acid is simply a combination of H^+ and a non-metallic anion that does not contain oxygen use the prefix **hydro-** and replace the **-ide** portion of the anionic species with **-ic acid**.

Acids With Oxygen in the Anions

Naming an acid with polyatomic anions that do contain oxygen requires knowing the name of the polyatomic ion. If the name of the polyatomic anion ends with -ate replace the -ate with -**ic acid**. If the name of the polyatomic anion ends with **-ite** replace the -ite with **-ous acid**.

Basic Compounds

Basic compounds are named as metal hydroxides. For example $Ca(OH)_2$ would simply be named *calcium hydroxide*.

1.4.2. Neutralization Reactions

It is important to remember that when we are talking about the pH of a solution we are by necessity talking about an **aqueous** solution.

It is important to note that the above chemical equation describes an equilibrium reaction. In other words, the reaction arrows point both right and left.

If an acidic solution containing a certain concentration of H^+ is added to a basic solution containing a certain concentration of OH^-, the H^+ and OH^- ions will react to form H_2O which is the right to left reaction in the above chemical equation. These ions will have been said to have **neutralized** each other. If the H^+ and OH^- concentrations of the two solutions being combined are identical, the solution will be absent of acid and base concentration (excepting the 10^{-14} molar concentrations of pure water) and will have a pH = 7.00.

If the H^+ and OH^- concentrations of the two solutions being combined are not identical the pH can be calculated by the remaining H^+ or OH^- concentration after neutralization has occurred.

1.4.3. *Buffer Solutions*

A ***buffer*** solution is a solution that resists changes in pH. This resistance is accomplished by a phenomenon known as the ***common-ion effect*** where more of an ion involved in the aqueous equilibrium of a solution is added from another source.

The common-ion effect is a prime example of ***Le Châtelier's Principle***.

Le Châtelier's Principle: If the equilibrium of a reaction is stressed a reaction will occur that relieves that stress. In pH buffer solutions the common-ion effect is between a weak acid and it's conjugate base or a weak base and it's conjugate acid.

If a strong acid is added to a solution of acetic acid the pH would rapidly decrease. If sodium acetate (CH_3COONa) were added to the solution the concentration of the conjugate base acetate ion ($CH\,COO^-$) would increase and neutralize the acid added to the solution. The solution would be said to be ***buff red*** against a decrease in pH.

1.4.4. *Hydrated and Anhydrous*

Some materials that are soluble in water crystallize from water with water incorporated in them, which can be driven off by heating. The crystals are called "hydrates" (Greek *hudōr*, "water") and the water "water of crystallization". Driving this off leaves the "anhydrous" form. For example, soda-ash, which is anhydrous, crystallizes from water as a hydrate ("washing soda").

1.5. Molecular Mass

The universe is made up of matter. The matter is anything that occupies space and has mass or weight. The matter is made up of fundamental particles called elements. To further understand chemistry the basic concepts are essential.

1.5.1. *Atom*

An Atom is the smallest particle of an element which can take part in chemical reactions. It may or may not be capable of independent existence.

Democritus (ca. 460–370 BC) is usually accredited as the originator of ***atomism***. Atomism is the theory that matter is made up of tiny particles called ***atoms***. This single concept is the basis of modern day atomic theory. Throughout history some have claimed that the theory was no more than a lucky guess. The fact that his theory aligns so well with the theories of the 19th and 20th century has led others to name Democritus as "the father of science".

In the early 19th century John Dalton (1766–1844) developed what is now considered the foundation of modern atomic theory. The table below lists his five postulates of atomic theory.

Table 1: Postulates of Atomic Theory

Dalton's Atomic Theory
1) Elements are made up of extremely small particles called atoms.
2) Atoms of a given element are identical in size, mass, and other properties.
3) Atoms cannot be created or destroyed.
4) Atoms of different elements combine in fixed whole number proportions to form molecules.
5) In chemical reactions atoms are separated and recombined into different molecules.

Ernest Rutherford (1871–1937) is credited with the discovery of sub-atomic particles by performing his famous gold foil experiment. He bombarded a thin gold foil with radioactive particles and noticed two things. First, that in some areas the radioactive particles was reflected. Second, that in other areas the radioactive particles passed right through the foil. From these observations he theorized that the gold was comprised of dense centers of mass but was mostly open space. He theorized that the dense centers of mass were sub-atomic particles. He reasoned that hydrogen, the lightest element, contained one of these particles that he called a ***proton***.

Joseph John (J.J.) Thomson (1856–1940) is credited with the discovery of the ***electron***. By working with cathode rays, Thomson noticed that the particles were well over a thousand times smaller than other subatomic particles. He also noticed that these particles generated the same amount of heat and had the same magnetic deflection no matter what material they came from. Thomson named these new fundamental particles ***corpuscles*** which was later changed to electrons.

James Chadwick (1891–1974) discovered a subatomic particle of approximately same mass as the proton that had a neutral charge. This had already been predicted and was named the ***neutron***. He later worked on the Manhattan project and helped in the development of the first atomic bombs.

History notwithstanding, here is a brief review of what we have learned. An atom is comprised of the following:

- Positively charged particles called ***protons***.

- Neutrally charged particles called **neutrons**.

- A **nucleus** containing the protons and neutrons.

- Negatively charged particles orbiting the nucleus known as **electrons**.

1.5.2. *The Periodic Table*

Early on during the development of atomic theory other researchers were observing and developing theories on the behavior of matter. That is to say they characterized different types of matter based on how their properties were alike or unalike.

Dimitri Mendeleev (1834–1907) is credited with developing the first viable periodic table of the elements. There were others that had developed tables that showed the properties of different elements. The genius of Mendeleev was that his table contained gaps where he predicted as yet undiscovered elements would later fill those gaps. Some of the elements he predicted were discovered during his lifetime. These discoveries vindicated Mendeleev's notion of periodic behavior and is now the very bedrock of modern day periodic tables.

The Periodic Table of the Elements

	I A	II A	III B	IV B	V B	VI B	VII B		VIII B		I B	II B	III A	IV A	V A	VI A	VII A	VIII A
1	1 **H** 1.008																	2 **He** 4.003
2	3 **Li** 6.941	4 **Be** 9.012											5 **B** 10.81	6 **C** 12.011	7 **N** 14.007	8 **O** 15.999	9 **F** 18.998	10 **Ne** 20.179
3	11 **Na** 22.990	12 **Mg** 24.305											13 **Al** 26.98	14 **Si** 28.09	15 **P** 30.97	16 **S** 32.07	17 **Cl** 35.45	18 **Ar** 39.95
4	19 **K** 39.10	20 **Ca** 40.08	21 **Sc** 44.96	22 **Ti** 47.57	23 **V** 50.94	24 **Cr** 52.00	25 **Mn** 54.94	26 **Fe** 55.85	27 **Co** 58.93	28 **Ni** 58.70	29 **Cu** 63.55	30 **Zn** 65.39	31 **Ga** 69.72	32 **Ge** 72.61	33 **As** 74.92	34 **Se** 78.96	35 **Br** 79.90	36 **Kr** 83.80
5	37 **Rb** 85.47	38 **Sr** 87.62	39 **Y** 88.91	40 **Zr** 91.22	41 **Nb** 92.91	42 **Mo** 95.94	43 **Tc** (98)	44 **Ru** 101.07	45 **Rh** 102.91	46 **Pd** 106.42	47 **Ag** 107.87	48 **Cd** 112.41	49 **In** 114.82	50 **Sn** 118.71	51 **Sb** 121.76	52 **Te** 127.60	53 **I** 126.90	54 **Xe** 131.29
6	55 **Cs** 132.91	56 **Ba** 137.33	57-71 **La**	72 **Hf** 178.49	73 **Ta** 180.95	74 **W** 183.85	75 **Re** 186.21	76 **Os** 190.23	77 **Ir** 192.22	78 **Pt** 195.08	79 **Au** 196.97	80 **Hg** 200.59	81 **Tl** 204.38	82 **Pb** 207.2	83 **Bi** 208.98	84 **Po** (209)	85 **At** (210)	86 **Rn** (222)
7	87 **Fr** (223)	88 **Ra** (226)	89-103 **Ac**	104 **Rf** (261)	105 **Db** (262)	106 **Sg** (263)	107 **Bh** (262)	108 **Hs** (264)	109 **Mt** (266)	110 **Uun** (264)	111 **Uuu** (272)	112 **Uub** (277)	113 **Uut** (284)	114 **Uuq** (289)	115 **Uup** (288)	116 **Uuh** (292)	117 **Uus**	118 **Uuo** (294)

Lanthanides	57 **La** 138.91	58 **Ce** 140.12	59 **Pr** 140.91	60 **Nd** 144.24	61 **Pm** (145)	62 **Sm** 150.36	63 **Eu** 151.96	64 **Gd** 157.25	65 **Tb** 158.93	66 **Dy** 162.50	67 **Ho** 164.93	68 **Er** 167.26	69 **Tm** 168.93	70 **Yb** 173.04	71 **Lu** 174.97
Actinides	89 **Ac** (227)	90 **Th** 232.04	91 **Pa** 231.04	92 **U** 238.03	93 **Np** (237)	94 **Pu** (244)	95 **Am** (243)	96 **Cm** (247)	97 **Bk** (247)	98 **Cf** (251)	99 **Es** (252)	100 **Fm** (257)	101 **Md** (258)	102 **No** (259)	103 **Lr** (262)

Figure 3: Periodic Table

This presentation of The Periodic Table has several main objectives.

- To distinguish between metals, non-metals, and semi-metals. Here, the metals are colored blue, the non-metals are colored yellow, and the semi-metals are colored orange.

- To enumerate the eighteen groups. The eighteen groups are divided between the Main Group elements, groups I A through VIII A, and the Transition Metals, groups I B through VIII B. Note that group VIII B is actually comprised of three vertical columns.
- To enumerate the seven periods which are the vertical column of numbers on the left hand side of the table.
- To list the number of protons, or Z number, the average mass, or A number, and the chemical symbol.
- The following figure shows an example of this presentation.

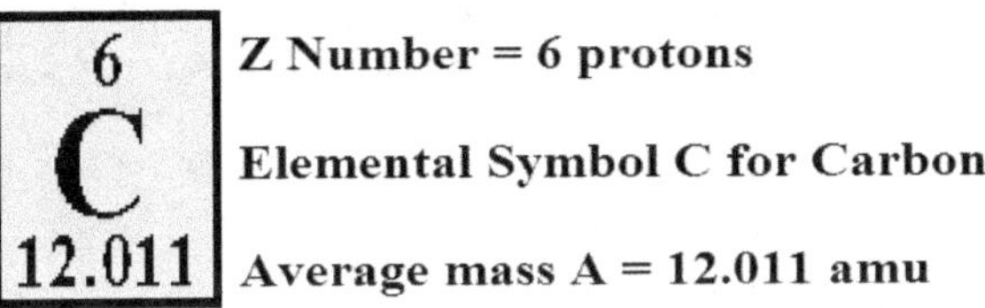

Figure 4: Representation of Atom

The nucleus of the carbon atom is comprised of six protons and six neutrons for a total of twelve **nucleons**. A neutron has essentially the same mass as a proton so that each nucleon is assigned a mass of 1 amu. The mass of an electron is 1/1800 that of a nucleon, therefore it's mass is not **significant**.

To calculate the number of neutrons in a carbon atom you must calculate the mass of a specific isotope. For example the number of neutrons in a carbon 12 atom would be the A number 12 minus the Z number 6 for a total of 6 neutrons.

So the number of neutrons in an element is as simple as this formula: **# neutrons = A–Z** (A expressed as the nearest whole number).

Here is a more specific look at what you should know about the Periodic Table.

- The vertical columns are known as **groups** and are populated by elements with similar properties.
- The horizontal rows are known as **periods** and are populated by elements that increase by one proton for each position.
- Groups IA through VIIIA are known as the **Main Group Elements**.
- Groups IB through VIIIB are known as the **Transition Elements**.
- The Lanthanide and Actinide series are known as the **Inner Transition**

1.6.   Elements

The color coordinated divisions on the Periodic Table represent the following.

The blue entries are *metals.* Metals have a shine or "metallic luster", conduct electricity, are malleable (hammered into sheets), and ductile (drawn into wire).

The yellow entries are *non-metals*. Overall non-metals are the complete opposite of metals. They are poor conductors of heat and, with the exception of the graphite form of carbon, are poor conductors of electricity.

The orange entries are *semi-metals*. Semi-metals properties lie between those of metals and nonmetals (as does their position on the periodic table). Silicon and germanium (semiconductors) are semi-metals.

1.7.   Electronic Structure of Atoms

Before we look at how atoms interact with one another, we must first know how their electrons are arranged about the nucleus of the atom.

Electrons are arranged about the nucleus of the atom in ever increasing energy levels.

These energy levels are known as *shells.* The shells are divided into *subshells* and each subshell is divided into *orbitals*.

These divisions of energy levels for electrons are exactly structured in a *periodic* manner.

The electron shells are numbered from 1 to infinity. However, it may be difficult to imagine an atom with an infinite number of electron shells.

The electron subshells are not numbered. They are given letter designations arising from the spectral lines that each subshell generates. They are labeled as follows:

s from the *sharp* spectral line

p from the *principal* spectral line

d from the *diffuse* spectral line

f from the *fundamental* spectral line

Understanding that each electron subshell contains a discrete number of electron orbitals and that each electron orbital can contain up to two electrons, the following table is a first step to understanding the electronic structure of atoms.

Table 2: Subshell of Atoms

Shell #	Subshell	Configuration	#ofOrbitals	Max.#e^{-s}
1	s	1s	1	2
2	s	2s	1	2
	p	2p	3	6
3	s	3s	1	2
	p	3p	3	6
	d	3d	5	10
4	s	4s	1	2
	p	4p	3	6
	d	4d	5	10
	f	4f	7	14

Let's explore filling electron shells across the Periodic Table with three simple rules.

Aufbau Principle: Electrons will fill the lower energy levels first and build up to the higher energy levels.

Pauli Exclusion Principle: Each electron orbital is limited to a maximum of two spin-opposed electrons.

Hund's Rule: Unfilled orbitals will be occupied before occupied orbitals are paired. (Fill orbitals so there is a minimum number of electron pairs.)

The following are examples of elements in each of the electron shells.

1.7.1. Electron Shell 1

This shell contains a single "s" subshell that in turn contains one electron orbital. This subshell, designated "1s", can contain 0, 1, or 2 electrons in its one orbital. The only two elements that contain only a 1s subshell are hydrogen and helium.

Note that the circle in my electronic structure diagrams represents an orbital. The half-barbed arrows represent electrons. In a full (two electron) orbital the two half-barbed arrows point in the opposite direction representing two spin-opposed electrons.

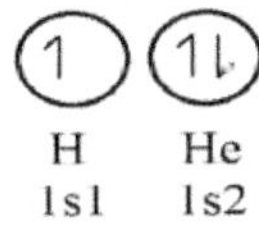

Figure 5: Electron Shell 1

1.7.2. *Electron Shell 2*

This shell contains a single "s" subshell and a single "p" subshell. The 2s subshell is the same as the 1s subshell. It contains one orbital that can contain 0, 1, or 2 electrons but at a higher energy level. The 2p subshell contains three orbitals that can contain 0, 1, or 2 electrons each for a total of six electrons.

Let's look at oxygen-8. Z=8 so there are 8 electrons. Place the electrons in the subshells and orbitals thus. 1s2 2s2 2p4

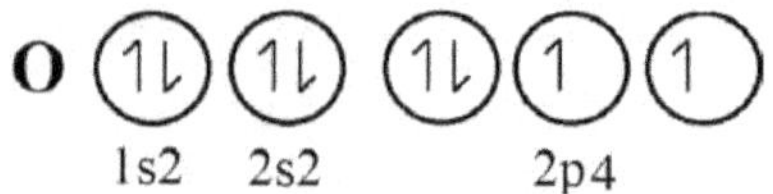

Figure 6: Electron Shell 2

Please notice that, **as per Hund's Rule**, 2p electron distribution forms as few electron pairs as possible.

1.7.3. *Electron Shell 3*

This electron shell contains single s, p and d subshells. The s and p subshells are identical to the previous s and p subshells. They have the same number of orbitals and electron capacity. The 3d subshell contains five orbitals that can contain from 0 to 10 electrons.

Let's look at iron. Z=26 so there are 26 electrons. Place the electrons in the subshells and orbitals thus. 1s2 2s2 2p6 3s2 3p6 4s2 3d6

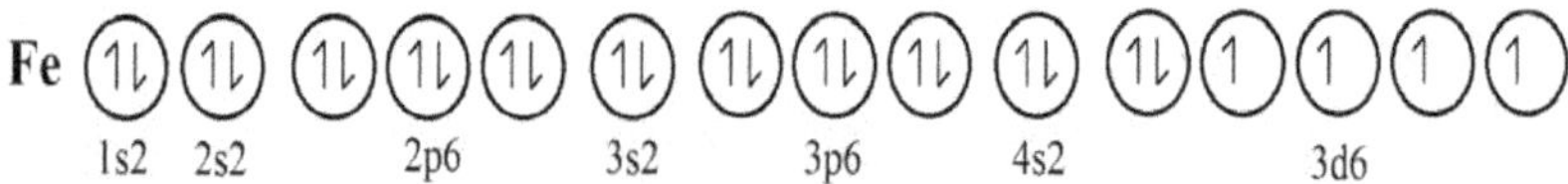

Figure 7: Electron Shell 3

It seems whenever Mother Nature sets up some elegant rules or relationships there are always exceptions. In filling of electron shells there are two important exceptions. The **next "s"** subshell will fi **before** the **present "d"** subshell begins to fill.

Except when a **half filled** or **fully filled** "d" subshell can exist. Let's look at the example of chromium (Z=24) and copper (Z=29).

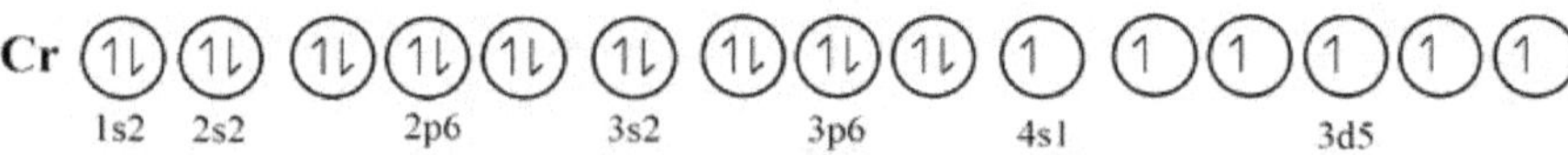

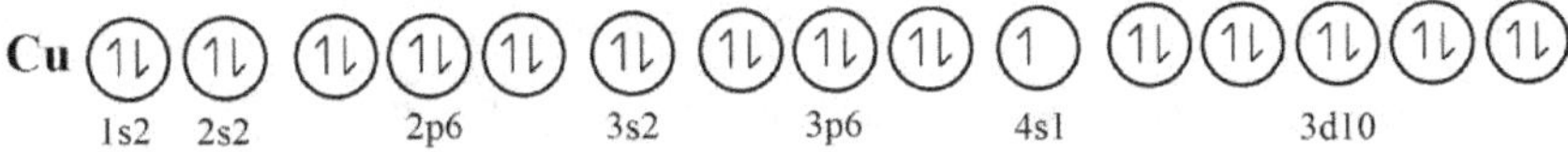

Here you see that the chromium is in a 4s1 3d5 configuration rather than a 4s2 3d4 configuration. Also, copper is in a 4s1 3d10 configuration rather than a 4s2 3d9 configuration. In both cases it is this way because the demonstrated configurations are at a lower energy level.

1.8.   Elements and their Isotopes

An element is listed as a unique entry on the Periodic Table due solely to the number of protons that element contains. As you proceed from the left to the right on any period of the Periodic Table the atomic number (Z) increases by one whole number for each new element. This is because the atomic number (Z) equals the number of protons.

Each element represented on the Periodic Table is represented as a neutral species. That is to say a species with a neutral charge. Therefore, the number of electrons for each element as it is represented on the Periodic Table is equal to the number of protons for that element.

Two atoms that contain the same number of protons (and therefore the same number of electrons) but different numbers of neutrons are known as *isotopes* of the same element with different atomic masses.

The figure below shows examples of a nucleus of pure carbon-12 containing 6 protons and 6 neutrons and a nucleus of carbon-14 containing 6 protons and 8 neutrons.

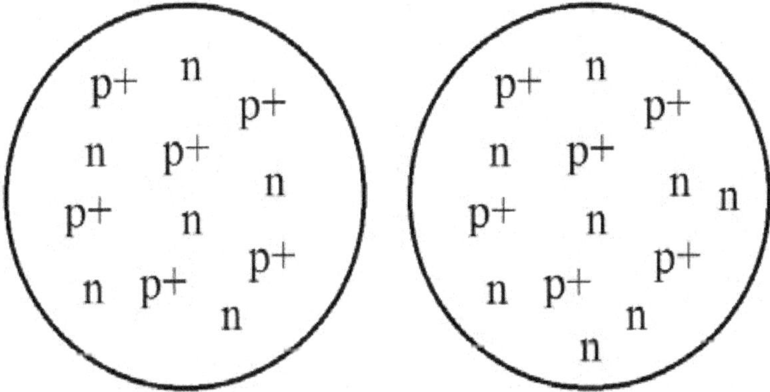

Figure 8: Two Isotopes of Carbon

Table 3: Average Atomic Mass

Z-Number	Element Name	Element Symbol	Average Atomic Mass
1	Hydrogen	H	1.008
2	Helium	He	4.003
3	Lithium	Li	6.941
4	Beryllium	Be	9.012
5	Boron	B	10.81
6	Carbon	C	12.011
7	Nitrogen	N	14.007
8	Oxygen	O	15.999
9	Fluorine	F	18.998
10	Neon	Ne	20.179
11	Sodium	Na	22.990
12	Magnesium	Mg	24.305
13	Aluminum	Al	26.98
14	Silicon	Si	28.09
15	Phosphorous	P	30.97
16	Sulphur	S	32.07
17	Chlorine	Cl	35.45
18	Argon	Ar	39.95
19	Potassium	K	39.10
20	Calcium	Ca	40.08
21	Scandium	Sc	44.96
22	Titanium	Ti	47.57
23	Vanadium	V	50.94
24	Chromium	Cr	52.00
25	Manganese	Mn	54.94
26	Iron	Fe	55.85
27	Cobalt	Co	58.93
28	Nickel	Ni	58.70
29	Copper	Cu	63.55
30	Zinc	Zn	65.39
31	Gallium	Ga	69.72
32	Germanium	Ge	72.61
33	Arsenic	As	74.92
34	Selenium	Se	78.96
35	Bromine	Br	79.90
36	Krypton	Kr	83.80

Z-Number	Element Name	Element Symbol	Average Atomic Mass
37	Rubidium	Rb	85.47
38	Strontium	Sr	87.62
39	Yttrium	Y	88.91
40	Zirconium	Zr	91.22
41	Niobium	Nb	92.91
42	Molybdenum	Mo	95.94
43	Technetium	Tc	(98)
44	Ruthenium	Ru	101.07
45	Rhodium	Rh	102.91
46	Palladium	Pd	106.42
47	Silver	Ag	107.87
48	Cadmium	Cd	112.41
49	Indium	In	114.82
50	Tin	Sn	118.71
51	Antimony	Sb	121.76
52	Tellurium	Te	127.60
53	Iodine	I	126.90
54	Xenon	Xe	131.29
55	Cesium	Cs	132.91
56	Barium	Ba	137.33
57	Lanthanium	La	138.91
58	Cerium	Ce	140.12
59	Praseodymium	Pr	140.91
60	Neodymium	Nd	144.24
61	Promethium	Pm	(145)
62	Samarium	Sm	150.36
63	Europium	Eu	151.96
64	Gadolinium	Gd	157.25
65	Terbium	Tb	158.93
66	Dysprosium	Dy	162.50
67	Holmium	Ho	164.93
68	Erbium	Er	167.26
69	Thulium	Tm	168.93
70	Ytterbium	Yb	173.04
71	Lutetium	Lu	174.97
72	Hafnium	Hf	178.49

Z-Number	Element Name	Element Symbol	Average Atomic Mass
73	Tantalum	Ta	180.95
74	Tungsten	W	183.85
75	Rhenium	Re	186.21
76	Osmium	Os	190.23
77	Iridium	Ir	192.22
78	Platinum	Pt	195.08
79	Gold	Au	196.97
80	Mercury	Hg	200.59
81	Thallium	Tl	204.38
82	Lead	Pb	207.2
83	Bismuth	Bi	208.98
84	Polonium	Po	(209)
85	Astatine	At	(210)
86	Radon	Rn	(222)
87	Francium	Fr	(223)
88	Radium	Ra	(226)
89	Actinium	Ac	(227)
90	Thorium	Th	232.04
91	Protactinium	Pr	140.91
92	Neodymium	Nd	144.24
93	Promethium	Pm	(145)
94	Plutonium	Pu	(244)
95	Americium	Am	(243)
96	Curium	Cm	(247)
97	Berkelium	Bk	(247)
98	Californium	Cf	(251)
99	Einsteinium	Es	(252)
100	Fermium	Fm	(257)
101	Mendelevium	Md	(258)
102	Nobelium	No	(259)
103	Lawrencium	Lr	(262)
104	Rutherfordium	Rf	(261)
105	Dubnium	Db	(262)
106	Seaborgium	Sg	(263)
107	Bohrium	Bh	(262)
108	Hassium	Hs	(264)
109	Meitnerium	Mt	(266)
110	Ununnilium	Uun	(264)
111	Unununium	Uuu	(272)
112	Ununbium	Uub	(277)
113	Ununtrium	Uut	(284)
114	Ununquadium	Uuq	(289)
115	Ununpentium	Uup	(288)
116	Ununhexium	Uuh	(292)
117	Ununseptium	Uus	
118	Ununoctium	Uuo	(294)

A Molecule is defined as the smallest particle of matter (element or compound) which can exist independently. It may be made up of two or more atoms of the same or different elements. For example Molecules of Chlorine, Oxygen, Nitrogen contain only two atoms. Molecules of Chlorine, Oxygen, Nitrogen contain only two atoms.

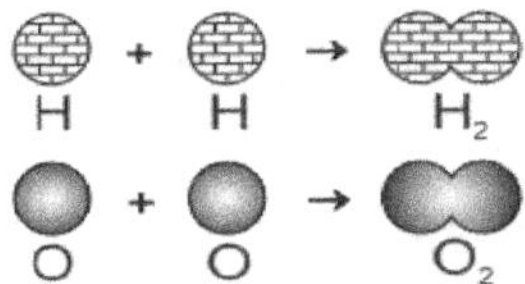

Figure 9: Molecule of H_2O

Molecules of sulphur dioxide, carbon dioxide etc., are built up by more than two atoms and involve the combinations of atoms of different elements. Thus in the molecule of carbon di oxide, one atom of carbon and two atoms of oxygen have united.

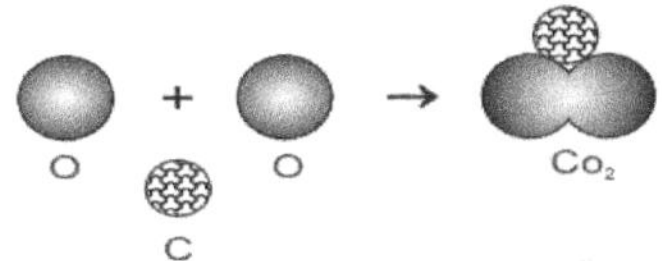

Figure 10: Molecule of C_2O

Whether matter can be subdivided continuously or is made up of atoms is a question that has intrigued thinkers back to the ancient Greeks if not before. Th e are several indications that matter may be made up of atoms.

1.9. Molecular Formula

Molecular formula is the short form of symbolic representation of one molecule of an element or a compound.

Example   : Molecular formula of Oxygen is O (element).

              : Molecular formula of Water is H O (compound)

1.9.1. *Significance of Molecular Formula*

1. Molecular formula represents one molecule of an element or a compound.
2. It shows the elements present in one molecule.
3. It gives the number of atoms of each element present in one molecule.
4. It helps to calculate the molecular mass.
5. It gives the ratio between the masses of the elements to form the substances.

1.10. Molecular Mass

Molecular mass of an element or a compound is the ratio between the mass of one molecule of the element or the compound and the mass of 1/12 part of a carbon Atom.

$$\text{Molecular Mass of an element or a compound} = \frac{\text{Mass of one molecule of an element or a compound}}{1/12 \text{ part by mass of a carbon atom}}$$

1.10.1. Calculation of Molecular Mass

Molecular mass can be calculated as the sum of total atomic mass of each element present in one molecule of an element or a compound.

Example 1

Molecular mass of O_2 = Atomic mass x No of atoms

$$= 16 \times 2 = 32$$

Molecular mass of NH_3 = $(14 \times 1) + (1 \times 3) = 17$

1.11. Mole

If the molecular mass is expressed in grams, then it is called gram molecular mass or mole.

$$\text{No of moles} = \frac{\text{Mass in gram}}{\text{Molecular mass}}$$

Example 2

Molecular mass of O = 32

Gram molecular mass of O_2 (or) 1 mole of O_2 = 32 g

Problem

1. How many moles are present in 8.5g of Ammonia?

Atomic mass of Nitrogen =14

Atomic mass of Hydrogen =1

Molecular mass of Ammonia (NH_3)=$(14 \times 1) (1 \times 3)= 17$

$$\text{No of moles} = \frac{\text{Mass}}{\text{Molecular mass}} = \frac{8.5}{17} = 0.5$$

Thus, no. of moles of NH_3= 0.5

1.12. Basic Laws and Phenomena of Chemistry

1.12.1. Quantitative Chemistry

In the history of chemistry, an understanding of chemical reactions came from measuring the quantities of the substances taking part in them, i.e. their masses (m) or, in the case of gases, volumes (V). The volume of a gas varies with temperature (t) and pressure (p), so these have to be specific The usual conditions are "standard temperature and pressure" (STP), = 0 ºC and p = 1 atmosphere (atm) (= 1.013 bar). Volumes can be converted to STP by invoking the gas laws

1.13. Properties of Gases

Units of Gas Measurements

Gases are unique in comparison to solids and liquids in that they do not have a fixed volume. Imagine that you are standing in a room holding a balloon. The balloon contains a certain volume of gas. If you pierce the balloon the volume of the gas then becomes the volume of the room. The following table shows the properties of gases that are measured to help characterize gases.

Table 4: Measurements of Gases

Measurements	Units	Symbols
Pressure	Pascal	Pa
Temperature	Kelvin	K
Volume	Liter	L
Amount	Moles	n

The Pascal is the SI unit for pressure where 1 Pa = $1N/m^2$. However, it is not uncommon to hear the pressure of gases referred to in different units. The following table lists some alternate units for the measurement of gas pressure.

Table 5: Alternate Units for Gas Pressure

Measurement	Symbol	Equivalence to 1 atm
mm of Mercury	mmHg	760 mmHg
Pounds per Square Inch	lb/in^2	$14.7 lb/in^2$
Torr	torr	760 torr
Pascal	Pa	101,325 Pa
Atmosphere	atm	1 atm

Example 3

Convert 745 mmHg to Pascal.

If all of the values in Table 5 are equivalent to 1 atmosphere then they are all equivalent to one another.

745 mmHg × 101,325 Pa / 760 mmHg = 99,325 Pa

Example 4

Convert 1800 torr to pounds per square inch.

1800 torr × 14.7 lb/in^2 / 760 torr = 34.8 lb/in^2

Example 5

Convert 450,000 Pa to atmospheres.

450,00 Pa × 1 atm / 101,325 Pa = 4.44 atm

1.14. The Kinetic-Molecular Gas Theory

Early researchers gathered a series of postulates concerning the nature of gases that are now known as the Kinetic-Molecular Gas Theory.

The following are the postulates of the Kinetic-Molecular Gas Theory:

- Gases are composed of a very large number of particles moving in random directions.
- The particles are so far apart from one another that their aggregate volume, as compared to the volume of the gas, is insignificant.
- The particles travel in straight lines unless they collide with another gas particle or the walls of the container.
- When collisions occur between gas particles or the container walls they always occur in an *elastic* manner.
- There are no attractive forces between different gas particles or the walls of the container.
- The average kinetic energy of a gas is proportional to the temperature of the gas measured in Kelvin.

1.15. Elastic and Inelastic Scattering

Elastic scattering is defined as a scattering of particles where the kinetic energy of a particle is always conserved. If you drop a ball on a flat hard surface you will notice that the ball loses energy on each bounce. This is *inelastic scattering*.

The molecules in a gas lose no energy when they strike another gas molecule. This is *elastic scattering*.

1.16. Basic Gas Laws

1.16.1. Boyle's Law

Boyle's Law states that volume is inversely proportional to pressure at a constant temperature and amount.

If you press a piston into a cylinder you increase pressure and the volume of the cylinder beneath the piston decreases.

At constant temperature in a closed system the relationship of changing volume and pressure is expressed as follows.

Where:

P is pressure

V is volume

1.16.2. Kinetic Theory

This theory explains some of the properties of materials. It is based on the idea that materials comprise a large number of small particles in continual motion. In a gas they move throughout the container. In a liquid they move throughout the liquid. In a solid they vibrate but do not move position.

The simplest case is that of a gas. Consider a gas in a cubic container of side L. Suppose that this comprises N particles of mass m. Let us further suppose, for simplicity, that all the particles are moving with the same speed s, and that $N/3$ are moving between the top and bottom face, $N/3$ between the left and right face, and $N/3$ between the front and back.

1.16.3. Avogadro Number

In the calculation on hydrogen peroxide in Section 6.2, I introduced the quantity g/u. This is a number – the ratio between two units of mass. It is called the "Avogadro number" (N_A), not to be confused with the "Avogadro constant" (L). It is approximately equal to the number of hydrogen atoms (mass ≈ 1 u) in one gram of hydrogen. It thus relates the atomic scale to the macroscopic. An estimate of its value can be obtained from the area (A) over which a measured volume (V) of oil spreads on water. The thickness of the oil (d) is given by V/A and, if the molecules of oil are cubes, their area is d^2. The number is therefore A/d^2

1.16.3.1. *Avogadro's Hypothesis*

Avogadro's Hypothesis states that "Equal volumes of all gases under the same conditions of temperature and pressure contain equal number of molecule.

Applications of Avogadro's Hypothesis

- It helps in deduction of the relationship between vapour density and molecular mass of gas.
- Avogadro's Hypothesis helps in determining the relationship between weight and volume of gases.
- It enables us to calculate the density of a gas.
- Avogadro's Hypothesis has established the truth about Dalton's atomic theory by making a clear distinction between atoms and molecules.
- Avogadro's Hypothesis helps in determining the atomicity of gases.
- It helps in the calculation of molecular mass of hydrocarbons.

According to Avogadro's hypothesis,

- One molecular mass (i.e. one mole) of every gas occupies 22.4 litres at S.T.P (or N.T.P)
- Equal volumes of all gases at S.T.P contain equal number of molecules. Hence it follows that 1 mole of every gas contains the same number of molecules. This number is called the Avogadro's number or Avogadro's Constant. It is denoted by. It has been found to be equal to 6.023 x 10. It can be defined as "the no of atoms or molecules present in one mole of an element or a compound respectively

Example 6

One mole of CO_2 = 44g

44 g of CO contains 6.023×10^{23} molecules.

One molar volume of CO_2 = 22.4 litres

22.4 litres of CO contains 6.023 x 10 molecules

Example 7

One mole of H_2O = 18g

18 g of H_2O contains 6.023 x 10 molecules

One mole of Oxygen atom = 16g

16 g of O_2 contains = 6.023 x 10 atoms

1.17. Chemical Bonding

When two atoms in a molecule strongly tend to remain together, they are said to be in chemical bonding with each other. In other words, it is said that a chemical bond has been established between the two atoms.

Definition

"A chemical bond may be defined as an attractive force which holds together the constituent atoms in a molecule" According to Kossel and G.N.Lewis (1916) who put forward the octet theory of valency, assumed that all atoms have a tendency to acquire a stable grouping of 2 or 8 valence electrons as the elements in the zero group (Noble gases). Thus it may be concluded that it is the tendency of the atoms to acquire a stable configuration or to complete their outermost orbit which is the cause of the chemical combination between them

1.17.1. Types of Bonding

- Ionic bond (or) Electrovalent bond or Polar bond
- Covalent bond or Non-Polar bond
- Co-ordinate covalent bond or Dative bond
- Metallic bond.

Ionic Bond

This type of bond is formed as a result of the complete transfer of one or more electrons from one atom to other. This bond is generally present in inorganic compounds

Example : Formation of Sodium Chloride

Covalent Bond

This type of bond is formed by the of a pair of electrons between two atoms. The shared electrons are contributed by the atoms. The covalent bond is indicated by a line called 'single bond.(-).' 22625 (2) Covalent bond: mutual sharing both

Example : Formation of Ammonia (NH)

Co-ordinate Covalent Bond

It is yet another type of linkage by virtue of which atoms acquire a stable configuration. Both transfer as well as sharing of electrons is involved in this mode of bond formation. The "shared pair" of electrons is supplied by one atom only and the other atom simply takes part in sharing. Thus "co-ordinate linkage is one in which the electron pair is contributed by one atom only and the sharing is done by both combining atoms". The atom which provides the shared pair of electrons (called lone pair) is termed the and the atom which accepts this pair for the purpose of forming the molecule is called the atom. The coordinate linkage is shown by an arrow mark ( ). The direction of the arrow points to the acceptor atom.

Example: Formation of Ammonium Ion [NH]

Metallic Bonding: (Electron Sea Model)

Metallic bond may be defined as the force which binds metal kernels to a number of electrons within its sphere of reference. (The atom without valence electrons is called kernal)

1.18. Hydrogen Bonding

One of the interesting factors critical to the understanding of chemistry and biochemistry is that of ***hydrogen bonding***.

Hydrogen bonding is a phenomenon that arises from the properties of polar molecules. It occurs when the hydrogen center of apparent charge "$\delta+$" in a polar molecule is attracted to the negative center of apparent charge "$\delta-$" in another polar molecule.

Using the same drawing of water as a polar molecule in Figure 11, it is easy to demonstrate a water molecule hydrogen bonding with another water molecule.

The following figure demonstrates one water molecule being ***weakly attracted*** to another water molecule as a result of hydrogen bonding.

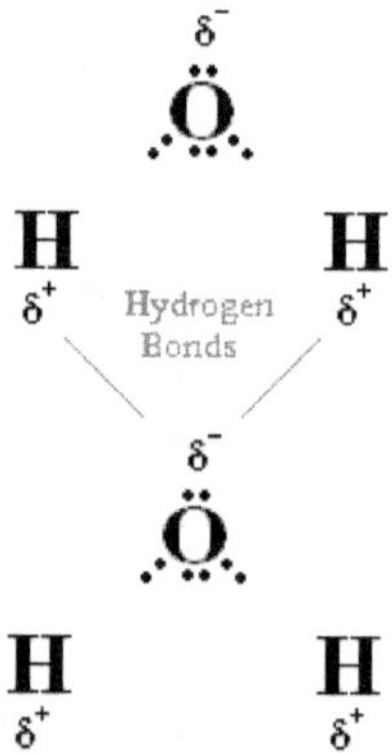

Figure 11: Hydrogen Bonding

One interesting result of hydrogen bonding in water is the effect it has on the boiling point of water. The following table contrasts the boiling points of several compounds, at or near 100°C, with the molecular masses of those compounds.

Table 6: Boiling Point Versus Molecular Mass

Compound	Formula	Mass (amu)	B.P.(°C)
Water	H_2O	18	100
n-Heptane	C_7H_{16}	100	98.4
iso-Octane	C_8H_{18}	114	99
Ethyl Acrylate	$C_5H_8O_2$	100	99.4

1.19. Chemical Reactions

1.19.1. Balancing Chemical Equations

In all chemical equations Dalton's original concept of the **conservation of mass** applies directly. That is to say the mass of the products **must** equal the mass of the reactants.

Later in the text we will learn that mass is not conserved in nuclear reactions.

1.19.2. Stages of Chemical Equations

Chemical equations, like the above metatheses reaction, can be expressed in different stages according to different needs.

The above metatheses reaction is stated in the stage known as the **complete chemical equation.** This stage shows all of the compounds present in the reaction in their original states.

$$2Na_3PO_4(aq) + 3MgCl_2(aq) \rightarrow Mg_3(PO_4)_2(s) + 6NaCl(aq)$$

Another stage of the same reaction is the **ionic equation.** In this stage the components of the reaction are shown as they exist in solution.

$$6Na^+ + 2PO_4^{3-} + 3Mg^{2+} + 6Cl^- \rightarrow Mg_3(PO_4)_2(s) + 6Na^+ + 6Cl^-$$

Please note that the magnesium phosphate ($Mg_3(PO_4)_2(s)$) is not in solution. It was precipitated in its non-soluble solid state.

A final and simpler stage of this metatheses reaction is known as the **net ionic equation.** In this stage of the chemical equation only the ionic portions of the original compounds that actually undergo metatheses, or a change, are listed.

If you notice, in the ionic equation of the metatheses reaction, there are $6Na^+$ and $6Cl^-$ on both sides of the reaction. They are not taking place in any metatheses and are known as **spectator ions.** It is not necessary to indicate an aqueous species for an ion showing either a positive or negative charge.

1.20. Oxidation (Redox) Reactions

One type of chemical reaction that uses net ionic equations are ***reduction-oxidation (redox)*** reactions. In a redox reaction a ***balanced transfer of electrons*** occurs that changes the oxidation state of some of the components of the reaction. Let's see some examples.

One can readily see that the iron (Fe) was changed from a 2+ oxidation state in the reactants to a 3+ oxidation state in the products.

One cannot readily see the change of the oxidation state of the manganese (Mn). The individual oxidation state of manganese in the reactants is hidden since the manganese exists in a polyatomic ion. Fortunately the polyatomic ion contains oxygen (O). Since oxygen is a Main Group element you already know it has a charge of 2- from Table 3.4.1.Set up an equation to solve for the oxidation state of manganese in the reactants as follows:

Mn + 4(2-) = -1 where the Mn charge can now be solved for. A simple calculation shows that manganese is in the 7+ oxidation state in the reactants. The final results of the redox reaction are as follows.

Iron went from 2+ to 3+ by a loss of 1 electron and was ***oxidized***. Manganese went from 7+ to 2+ by a gain of 5 electrons and was ***reduced***. It gained 1 electron each from the 5 irons.

1.20.1. Leo the Lion

A very effective mnemonic device used to remember the direction of electron transfer in redox reactions is ***LEO the lion goes GER***.

A **L**oss of **E**lectrons is **O**xidation. A **G**ain of **E**lectrons is education.

Some Reasoning Concerning Redox Reactions

The simple arithmetic manipulations shown in the previous examples and the Leo the Lion mnemonic device are very easy to follow. But some sense of what is happening in a redox reaction is in order.

Here are four important items to remember when considering redox reactions.

- Reduction and oxidation happens as a single reaction. You ***cannot*** have one without the other.
- The species that is becoming oxidized is reducing the other species and is the ***reducing agent.***
- The species that is becoming reduced is oxidizing the other species and is the ***oxidizing agent.***

- Close inspection of the redox equation will **always** show a balanced transfer of electrons.

1.21. The Concept of the Mole

Chemistry students often struggle with the concept of a **mole**. By definition a mole is 6.022×10^{23} of **anything**. It is just another number with a name. Reflect on some common numbers with names and then simply include the mole.

1.21.1. Molar Ratios

It is not an oversimplification to state that the number of moles in a compound occur at the exact same ratios as the number atoms in a molecule. The subscripts and coefficients in chemical formulae represent the number of atoms in the molecule and the number of moles of atoms in a mole of the molecule.

When reading the molecular formula for a certain compound you can now use coefficients and subscripts to assess the number of **moles of atoms** of a certain atom in **one mole of the compound**. Some examples will clarify this point.

Example8

How many moles of H are in one mole of H2O?

If there are 2 H atoms in each H_2O molecule then there are 2 moles of H in each mole of H_2O.

Example9

How many moles of H are there in 5 moles of C_3H_8 (propane)?

If there are 8 H atoms in each C_3H_8 molecule then there are 8 moles of H in one mole of C_3H_8 and 40 moles H in 5 moles C_3H_8.

Example10

How many moles of H are there in 1 moles of $CO(NH_2)_2$ (urea)?

Note that here there are **three multipliers** for H in the chemical formula $1CO(NH_2)_2$. You must multiply H by the coefficient 1, the 2 inside the parentheses, and the 2 outside the parentheses.

The single H in the formula multiplied by $1 \times 2 \times 2 = 4$. There are 4 moles of H in 1 mole of $CO(NH_2)_2$.

To make the concept of molar ratios a little clearer, let's look at the number of H atoms in a drawing of the $CO(NH_2)_2$ molecule. Remember that this number is also the number of moles of H in 1 mole of $CO(NH_2)_2$.

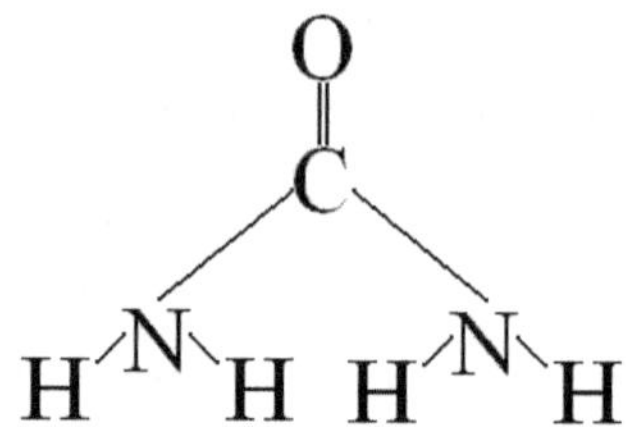

Figure 12: The Urea Molecule

It is easy to see the 4 H atoms in the drawing of urea in the above figure. This number **must also** be interpreted as the number of moles of H in 1 mole of $CO(NH_2)_2$.

1.22. Stoichiometric Calculations

Stoichiometry is the branch of chemistry that deals with the calculations of amounts of reactants consumed and products formed in a chemical reaction. The methods used to make these calculations have been being developed all along in the progression of this text.

Decimal Forms of Coefficients

Consider the following balanced chemical equation for the combustion of butane (C_4H_{10}).

$$C_4H_{10} + 6.5O_2 \rightarrow 4CO_2 + 5H_2O$$

It is **impossible** to have 6.5 molecules of O_2 in the equation of this chemical reaction. However, it is **perfectly acceptable** to have 6.5 moles of O_2 in the equation for this reaction.

1.23. Types of Homogeneous Mixtures

There are basically three types of **homogeneous mixtures.** This means that the components of the mixtures are evenly dispersed throughout the volume of the mixture.

Solutions

Solutions contain particles in the range of tenths of nanometers (nm) to several nm. A nanometer is 10^{-9} meters. They are transparent and, though they may be colored, do not separate on standing.

Colloids

Colloids, such as homogenized milk and fog, contain particles in the range of hundreds of nanometers. Although they are often murky or opaque to light, they do not separate on standing.

Suspensions

Suspensions are heterogeneous, nonuniform mixtures with particles so large they can be separated by use of filters or semi-permeable membranes.

Solution Nomenclature

It is important when discussing solutions that one use the agreed upon terminology that describes the constituents of a solution. Listed below are three very ***important*** definitions in solution chemistry.

Solution

A homogeneous mixture of molecules or ions.

Solvent

The medium in which the molecules or ions are dissolved.

Solute

Any substance dissolved in a solvent.

Concentration

The ratio of the amount of solute to the amount of solution. Although we will deal entirely with liquid solutions in this chapter, it should be noted that solutions do exist in all of the three states of matter. The following table shows examples of solutions in liquid, gaseous, and solid states of matter.

Table 7: Solutions of Different States of Matter

State of Matter	Solvent	Solute	Solution
Liquid	Water	Acetic Acid	Vinegar
Gas	Nitrogen	Oxygen	Air
Solid	Copper	Tin	Bronze

1.24. Solution Properties

As you may have already noticed, chemistry, like other sciences, can be viewed as a systematic way of observing a component or system of nature. These observations are then recorded, agreed upon, and set down as the ***properties*** of that component or system.

Listed below are the properties of solutions that you will need to understand the chemistry of solutions.

Solubility

Solubility is the measure of how much of a solute will dissolve in a given amount of solvent. When solubility is measured in g/100 ml the 100 ml is the **given amount**. It has been agreed upon among chemists to measure and record solubilities in this manner using this specific unit.

For example, common table salt (NaCl), has a solubility in water of 39.5 g/100 ml. You could prove this to yourself by measuring out 100 ml of water into a clear glass. Measure out 50 grams of table salt and try to get all of the salt into solution in the 100 ml of water.

The table below lists solubilities of common substances in water at 20°C.

Table 8: Solubilities in Water at 20°C

Substance	Formula	Solubility (g/100ml)
Sucrose	$C_{12}H_{22}O_{11}$	200
Potassium Nitrate	KNO_3	47
Sodium Chloride	NaCl	39.5
Sodium Sulfide	Na_2S	1.86
Calcium Hydroxide	$Ca(OH)_2$	0.017

Saturation

Saturation is that state of a solution where the amount of solute added to the solvent has exceeded the solubility of that solute. When excess solute is added to a solvent, that solute will not go into solution. It is common to stir a solution to get solutes into solution. When excess solute is added in this manner, the solution will appear turbid at first due to suspended solutes but will turn clear when the excess solute settles to bottom of the container.

Super Saturation

Super saturation is the state of a solution when more solute is in solution than the solubility would allow for due to special circumstances. A sealed bottle of soda is a solution that is supersaturated with carbon dioxide (CO_2) due to the high pressure inside the bottle. When the bottle cap is removed the pressure rapidly decreases releasing the excess CO_2 to the atmosphere.

A rain cloud can be supersaturated with water vapor. In order for the water vapor to coalesce as liquid water it must be supplied with a non-gaseous surface to act as a point of nucleation. Once particles of dust are provided as "cloud seeds" the excess water vapor will quickly coalesce into rain drops.

Miscibility

Miscibility is the property of liquids to combine with one another in all proportions to form homogeneous solutions. In a manner of speaking miscibility is unbounded solubility. In table 8 I showed the solubilities of some ionic compounds and a sugar. Had I included acetic acid I would have listed the solubility as *miscible* since it will combine with water at all proportions.

At a range of 4% to 8% acetic acid in water the solution is commonly referred to as *vinegar*.

Concentration

Concentration is the measure of the amount of solute actually dissolved in a solvent. If you were to return to the experiment of adding 50 g of NaCl to 100 ml of water, under ideal conditions (pure water at 1 atm pressure and 20°C) the *concentration* of the NaCl in the water would be 39.5 g/100 ml. This is due to the limit of the solubility of NaCl in water.

1.25. Stages of Chemical Equations

Chemical equations, like the above metatheses reaction, can be expressed in different stages according to different needs. The above metatheses reaction is stated in the stage known as the *complete chemical equation.* This stage shows all of the compounds present in the reaction in their original states.

$$2Na_3PO_4(aq) + 3MgCl_2(aq) + 6NaCl(aq)$$

Another stage of the same reaction is the *ionic equation*. In this stage the components of the reaction are shown as they exist in solution.

$$6Na^+ + 2PO_4^{3-} + 3Mg_3^{2+} + 6Na^+ + 6Cl^-$$

Please note that the magnesium phosphate ($Mg_3(PO_4)_2(s)$) is not in solution. It was *precipitated* in its non-soluble solid state.

A final and simpler stage of this metatheses reaction is known as the *net ionic equation*. In this stage of the chemical equation only the ionic portions of the original compounds that actually undergo metatheses, or a change, are listed.

If you notice, in the ionic equation of the metatheses reaction, there are $6Na^+$ and $6Cl^-$ on both sides of the reaction. They are not taking place in any metatheses and are known as *spectator ions.* It is not necessary to indicate an aqueous species for an ion showing either a positive or negative charge.

1.26. Reduction-Oxidation (Redox) Reactions

One type of chemical reaction that uses net ionic equations are ***reduction-oxidation (redox)*** reactions. In a redox reaction a ***balanced transfer of electrons*** occurs that changes the oxidation state of some of the components of the reaction. Let's see some examples.

One can readily see that the iron (Fe) was changed from a 2+ oxidation state in the reactants to a 3+ oxidation state in the products.

One cannot readily see the change of the oxidation state of the manganese (Mn). The individual oxidation state of manganese in the reactants is hidden since the manganese exists in a polyatomic ion. Fortunately the polyatomic ion contains oxygen (O).Set up an equation to solve for the oxidation state of manganese in the reactants as follows:

Mn + 4(2-) = -1 where the Mn charge can now be solved for. A simple calculation shows that manganese is in the 7+ oxidation state in the reactants.

Iron went from 2+ to 3+ by a loss of 1 electron and was ***oxidized.***Manganese went from 7+ to 2+ by a gain of 5 electrons and was ***reduced***. It gained 1 electron each from the 5 irons. Staying within the SI system the 39.5 g/100 ml solubility of NaCl in water could be written in other terms of concentration. The 39.5 g/100 ml could be expressed as 395 g/L, 395 g/kg, 395,000 mg/kg, 395,000 ppm, and***6.76moles/L.***

1.27. Molarity

In Chapter 4 you were introduced to the concept of the mole. Like other measurements of amount you can convert the mole to concentration by expressing the number of moles per unit volume.

Molarity is the measure of the concentration of a solute in a solvent expressed in terms of the number of moles of that solute in one liter of the solvent.

1.27.1. Symbols and Terms of Molarity

It is important to become familiar with the different symbols and terms used to express molarity. The symbol for molarity is the upper case letter ***M***. However, there are a few different ways you may hear or see molarity expressed. Here are some examples of the way a solution that contains 1.5×10^{-2} moles of NaCl in one liter of water may be expressed:

- The solution is 1.5×10^{-2} *M* in NaCl.

- The solution has 1.5×10^{-2} *moles/liter* NaCl.

- This is a 1.5×10^{-2} *molar* solution of NaCl.

Just remember that they all mean the number of moles of solute per liter of solvent.

1.27.2. Molarity Calculations

Calculating molarity is a critical skill for many scientific disciplines. Climate scientists will use molarity calculations to estimate the load of carbon being deposited to the Earth's atmosphere.

Pharmaceutical chemists definitely use molarity calculations to determine the correct amounts of the various constituents of a drug product. A nurse could use molarity calculations to determine correct dosage regimens of medicines.

Exercise 11

What is the molarity of Cl^- in a 1 L solution of water containing 3 moles of NaCl?

$$\frac{3 \text{ moles NaCl}}{1 \text{ L}} = \frac{\text{moles } Cl^-}{L}$$

$$\frac{3 \text{ moles NaCl}}{1 \text{ L}} \cdot \frac{1 \text{ mole } Cl^-}{1 \text{ mole NaCl}} = 3 \frac{\text{moles } Cl^-}{L}$$

Here the molar ratio of Cl^- to NaCl is taken directly from the chemical formula.

Exercise 12

What is the molarity of OH^- in a 1.5 L solution of water that contains 4 moles of $Ca(OH)_2$?

$$\frac{4 \text{ moles } Ca(OH)_2}{1.5 \text{ L}} = \frac{\text{moles } OH^-}{L}$$

$$\frac{4 \text{ moles } Ca(OH)_2}{1.5 \text{ L}} \cdot \frac{2 \text{ moles } OH^-}{1 \text{ mole } Ca(OH)_2} = 5.3 \frac{\text{moles } OH^-}{L}$$

Here the chemical formula shows that there are 2 moles OH^- for 1 mole of $Ca(OH)_2$.

Exercise13

What is the molarity of Na^+ in a 0.75 L solution containing 0.35 g of NaBr?

$$\frac{0.35g\ NaBr}{0.75\ L} = \frac{moles\ Na^+}{L}$$

Convert g NaBr to moles NaBr (as shown below) using the periodic table.

The 1 mole of Na^+ in the 1 mole of NaBr has a molar mass of 22.99 g/mol. The 1 mole of Br- in the 1 mole of NaBr has a molar mass of 79.90 g/mol. Adding them together yields a molar mass for NaBr of 102.9 g/mol.

$$\frac{0.35g\ \cancel{NaBr}}{0.75\ L} \cdot \frac{1\ mole\ NaBr}{102.89\ g\ \cancel{NaBr}} = \frac{moles\ Na^+}{L}$$

Once again the molar ratio of Na^+ to NaBr comes directly from the chemical formula.

$$\frac{0.35g\ \cancel{NaBr}}{0.75\ L} \cdot \frac{1\ \cancel{mole\ NaBr}}{102.9\ g\ \cancel{NaBr}} \cdot \frac{1\ mole\ Na^+}{1\ \cancel{mole\ NaBr}} = 4.5 \times 10^{-3}\ \frac{moles\ Na^+}{L}$$

1.28. Mole Fractions

The Concept of Mole Fraction

Now that we know about both moles and molarity we can embrace the concept of mole fraction.

$$Mole\ Fraction\ (X) = \frac{moles\ of\ solute\ or\ solvent}{moles\ of\ solution}$$

A mole fraction can be expressed for either the solute or the solvent. Mole fraction is expressed as a fraction without units. If a certain amount of salt (NaCl) is added to a certain volume of pure water (H_2O), you should be able to calculate to mole fraction of Na^+ ions in that solution. You will need to calculate the number of moles of Na^+ ions, Cl^- ions, and water molecules that exist in that solution. Then the number of moles of Na+ ions divided by the number of moles of **all other constituents** in that solution is the mole fraction of Na+ ions in that solution.

Exercise14

What is the mole fraction of Na^+ ions in a 1 liter solution of pure water that is 0.1 molar in NaCl? Pure water has a density of 1kg/liter.

$$\frac{1L\ H_2O}{}\left|\frac{1kg\ H_2O}{1L\ H_2O}\right|\frac{1000g}{1kg}\left|\frac{1\ mol\ H_2O}{18g\ H_2O}\right. = 55.56\ mol\ H_2O$$

Now that we have calculated the number of moles we need to calculate the number of moles Na^+ and Cl^- ions.

$$\frac{0.1\ mol\ NaCl}{L}\left|\frac{1\ mol\ Na^+}{1\ mol\ NaCl}\right. = 0.1\ mol\ Na^+$$

Since Na^+ and Cl^- exist in an equal molar ratio in NaCl there are also be 0.1 moles of Cl^- in NaCl. Therefore in the solution there is a total of 55.56 + 0.1 + 0.1 = 55.76 of moles of all constituents.

The mole fraction of Na^+ ions in the solution is 0.1/ 55.76 or *1/557.6*.

Exercise15

What is the mole fraction of Cl^- in a 1 liter solution of pure water that has a 1×10^{-2} molar concentration of $CaCl_2$? From Exercise 6.5.1 we see that 1 liter of water contains 55.56 moles of water.

$$\frac{0.01\ mol\ CaCl_2}{}\left|\frac{2\ mol\ Cl^-}{1\ mol\ CaCl_2}\right. = \frac{0.02\ mol\ Cl^-}{L}$$

If there are 0.02 moles of Cl^- in the solution the 1 to 2 molar ratio of Ca^{2+} to Cl^- means there are 0.01 moles of Ca^{2+} in the solution. So there are a total of 55.56 + 0.02 + 0.01 = 55.59 moles of all constituents in the solution.

The Cl- mole fraction is 0.02/55.59 = *1/2795*.

1.29. Colligative Properties

Colligative properties are those properties of a solution that are dependent on the amount of a dissolved solute but not on the chemical identity of the solute. The following four colligative properties will be discussed:

- Vapor Pressure Depression Boiling Point Elevation
- Freezing Point Depression Osmotic Pressure

1.29.1. Vapor Pressure Depression

If the vapor pressure of a solvent is measured and compared with the vapor pressure of the same solvent, under identical conditions, to which a solute has been dissolved *the vapor pressure is lower.*

- ***Raoult's Law*** states that the vapor pressure of a solution containing a non-volatile solute is equal to the vapor pressure of the pure solvent times the mole fraction of the solvent.

Boiling Point Elevation

If the vapor pressure of a liquid has been lowered you will need to add more energy to raise the vapor pressure to atmospheric pressure. Therefore, the liquid will ***boil at a higher temperature.***

Freezing Point Depression

If a non-volatile solute is dissolved in a pure solvent the freezing point of the solution ***will be lower*** than the temperature of the pure solvent.

Hint: Think of the salt added to an ice-cream freezer.

Osmotic Pressure

If a vessel contains pure solvent and a solution of the solvent and a non-volatile solute separated by a semi permeable membrane, solvent molecules will migrate through the membrane in both directions. This is the process known as ***osmosis.*** Osmotic flow is favored toward the solution side of the vessel according to the ***osmotic pressure (Π)***

Osmosis in Blood Cell Biology

Because the cell membranes in biological systems are semipermeable, osmosis is an ongoing process. Most intravenous solutions are ***isotonic*** solutions because their osmotic pressure is in equilibrium with the pressure of the body fluids.

For instance, if an isotonic blood cell was placed in pure water, the resulting solution would be of a lower, or ***hypotonic,*** pressure and water would flow into the cell due to osmosis. This could cause the cell to rupture in a process known as ***hemolysis.***

In a concentrated solution, the same blood cell would realize a higher, or ***hypertonic,*** pressure solution resulting in fluid flowing out of the cell causing cell shrinkage in a process known as ***crenation.***

Fractional Distillation

Fractional distillation is a process of separating different components of a gas or liquid due to the fact that these components have different vapor pressures.

The most prevalent use for fractional distillation is the refinement of petroleum products.

Oil in an oil well consists of tens of thousands of different components. Most of these are liquid but many are in a vapor phase above the liquid.

Both the liquids and the gases are extracted and then fractionally distilled to produce the many product streams of the petroleum industry.

These products can vary from light hydrocarbons containing one or two carbons to heavy asphaltenes containing hundreds of carbons per compound.

Questions

Part A

1. What is Matter?
2. Define molecular mass of a substance.
3. Define mole
4. What is molecular formula?
5. State Avogadro's hypothesis.
6. Define Avogadro's number.
7. What are the types of bonding?
8. Define ionic bond.
9. Define covalent bond.
10. Define co-ordinate covalent bond.
11. What is metallic bond?
12. What are Types of Properties of Matter?
13. Define Element.
14. Define acids.
15. Define base.
16. Define salt.
17. Define an Atom.

Part B

1. Briefly Explain Basic Laws and Phenomena of Chemistry?
2. Briefly Explain Gas Law?
3. Explain in detail about Electronic Structure of Atom?
4. Define Isotope. Explain in detail?

CHAPTER 2

BASIC ORGANIC CHEMISTRY

2.1. Organic Molecules

Organic molecules are those molecules made up primarily of carbon and hydrogen.

Known as *hydrocarbons*, only a molecule consisting exclusively of carbon and hydrogen would be considered a pure hydrocarbon.

Methane (or natural gas) is the simplest hydrocarbon.

$$\textbf{Methane} \qquad H\!-\!\underset{\underset{\displaystyle H}{|}}{\overset{\overset{\displaystyle H}{|}}{C}}\!-\!H$$

Heteroatomic Hydrocarbons

Other atoms are formed along with carbon and hydrogen into organic molecules. Examples of these would be nitrogen (N), oxygen (O), sulfur (S), phosphorus (P), and others. These non-carbon, non-hydrogen atoms are then referred to as **heteroatoms**. Formaldehyde is a good example of a heteroatomic hydrocarbon.

$$\textbf{Formaldehyde} \qquad H\!-\!\overset{\overset{\displaystyle O}{\|}}{C}\!-\!H$$

2.2. Organic Compound Families

The organic compounds presented in this chapter will be divided into two major groups.

Pure hydrocarbons and their different families

For these compounds you will be expected to know **family names, structures, and carbon to hydrogen ratio formulae.**

Heteroatomic hydrocarbons and their different families

For these compounds you will be expected to know **family names and functional groups.**

2.3. The Alkanes

The first and the simplest of the pure hydrocarbons are the *alkanes*.

The alkanes are characterized by all single carbon to carbon covalent bonds.

The restriction to only single carbon to carbon bonds in a pure hydrocarbon results in alkanes having the maximum number of hydrogen atoms attached.

When a hydrocarbon has the maximum number of hydrogen atoms attached it is said to be saturated with hydrogen.

Examples of Alkanes

Here are shown the alkanes with the carbon numbers 1,2,3, and 4.

C1,C2, and C3 hydrocarbons cannot have isomers. The C4 hydrocarbon has one isomer. The carbon to hydrogen ratio formula for alkanes is C_nH_{2n+2}.

Figure 2.1 shows structures and formulae for the first four alkanes.

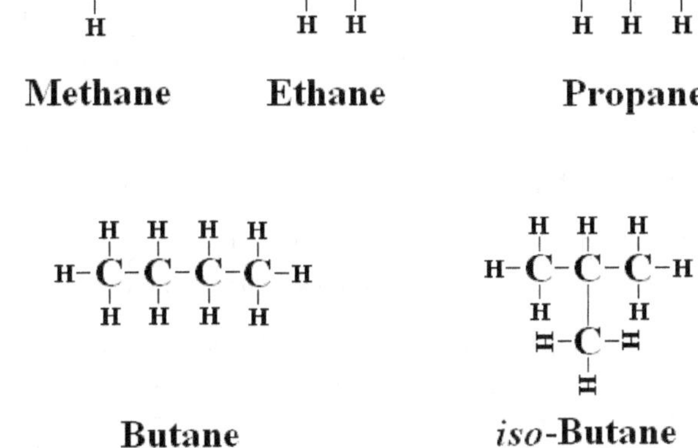

Figure 2.1: The C1, C2, C3, and C4 Alkanes

Notice that even though butane and *iso*-butane are different compounds they both have the same formula of C_4H_{10}.

They are structural isomers of one another.

Every alkane from butane up will have an increasing number of isomers of the straight chained version of the alkane.

Decane, the C10 alkane, has 75 structural isomers.

Instruction in this chapter is fairly well limited to compounds with ten or less carbon atoms.

2.3.1. Naming Alkanes

We will deal with alkanes that contain no more than 10 carbon atoms. Know these empirical formulae and names.

Table shows the formulae and names for the first ten unbranched alkanes.

C#	Structure	Name
1	CH_4	Methane
2	CH_3CH_3	Ethane
3	$CH_3CH_2CH_3$	Propane
4	$CH_3CH_2CH_2CH_3$	Butane
5	$CH_3CH_2CH_2CH_2CH_3$	Pentane
6	$CH_3CH_2CH_2CH_2CH_2CH_3$	Hexane
7	$CH_3CH_2CH_2CH_2CH_2CH_2CH_3$	Heptane
8	$CH_3CH_2CH_2CH_2CH_2CH_2CH_2CH_3$	Octane
9	$CH_3CH_2CH_2CH_2CH_2CH_2CH_2CH_2CH_3$	Nonane
10	$CH_3CH_2CH_2CH_2CH_2CH_2CH_2CH_2CH_2CH_3$	Decane

The alkanes listed in above table are all straight chained alkanes. That is to say there are no branches in their structures. These type of alkanes are referred to as **normal alkanes**.

The symbol for this is the lower case "n". If you were referring to pentane and wanted to be sure you stressed the straight chained isomer you would refer to it as **n-pentane**.

Alln-alkanes from butane up have branched isomers. These compounds are referred to as **branched alkanes**. Using "hydrogen suppressed" drawings the following diagrams show the three isomers of pentane (C_5H_{12}).

n-Pentane   2-Methylbutane   2,2-Dimethylpropane

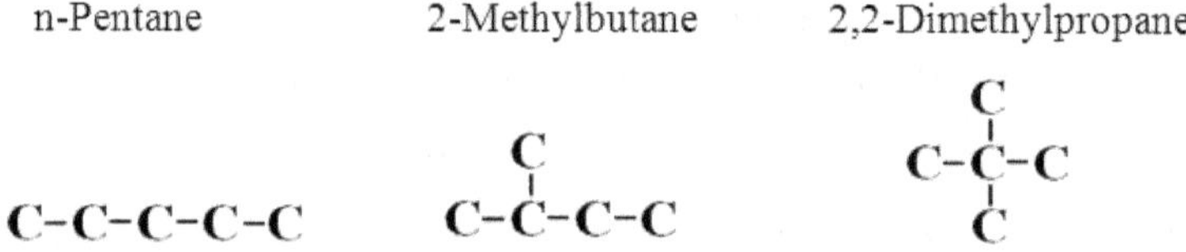

Figure 2.2: The Three Isomers of Pentane

The following rules were applied to naming the pentane isomers in Figure 2.2.

1. Each name was written as a continuous word and contained the name of the longest chain.

2. Each substituent was named by replacing the -ane of the alkane name with -yl.

3. The number of each substituent was assigned by numbering the longest chain to result in substituents with the smallest numbers.

4. Multiple substituents were named using the Greek prefixes.

Please note the two drawings of 2-methylbutane below.

Both of these drawings are of the same isomer of pentane and are numbered correctly.

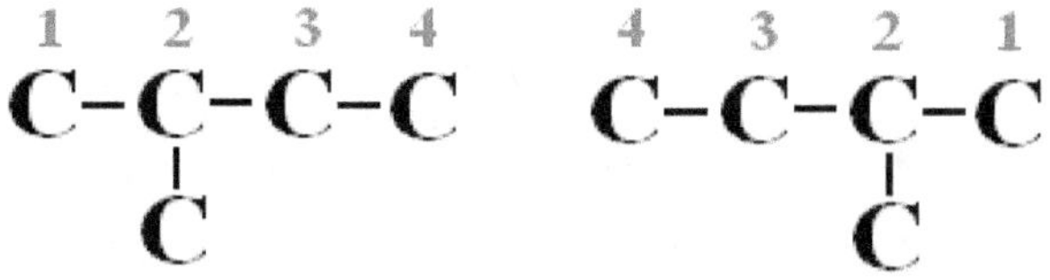

$$\overset{1}{C}-\overset{2}{\underset{\underset{C}{|}}{C}}-\overset{3}{C}-\overset{4}{C} \qquad \overset{4}{C}-\overset{3}{C}-\overset{2}{\underset{\underset{C}{|}}{C}}-\overset{1}{C}$$

Just because an isomer is drawn to the opposite hand does not make it a unique isomer.

Using the four rules listed previously and adding a fifth rule, the following branched alkane can be named.

5. Different substituents on an alkane chain appear in the name alphabetically.

Figure 2.3 shows 4-ethyl-3-methylheptane since "e" comes before "m" in the alphabet.

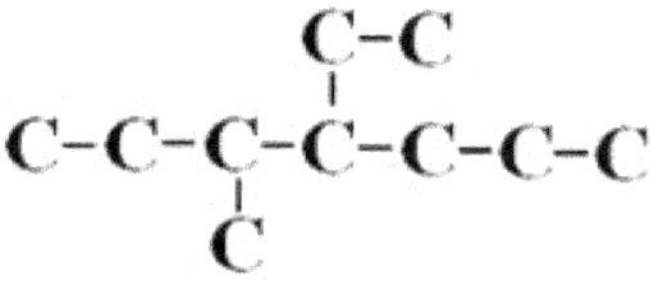

$$\begin{array}{c} C-C \\ | \\ C-C-\underset{\underset{C}{|}}{C}-C-C-C-C \end{array}$$

Figure 2.3: 4-Ethyl-3-methylheptane

2.3.2. Cycloalkanes

If you can imagine bringing the two ends of hexane together to form a ring, do so with one important rule in mind:

The Tetravalent Rule of Carbon

Carbon must always be bonded four times in a neutral molecule.

By following the tetravalent rule of carbon two hydrogens were given up in order to form the ring.

Figure 2.4 shows the cyclohexane molecule.

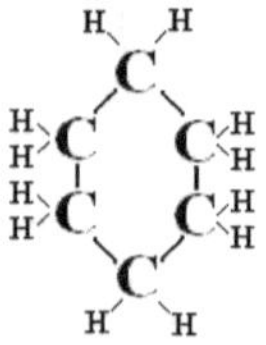

$$C_nH_{2n}$$

Figure 2.4: Cyclohexane

Those two hydrogens are known as 1 point of unsaturation.

2.3.3. A Review of Rules for Naming Alkanes

1. Each name was written as a continuous word and contained the name of the longest chain.

2. Each substituent was named by replacing the **-ane** of the alkane name with **-yl**.

3. The number of each substituent was assigned by numbering the longest chain to result in substituents with the smallest numbers.

4. Multiple substituents were named using the Greek prefixes.

5. Different substituents on an alkane chain appear in the name alphabetically.

6. Use the prefix "cyclo" to indicate a non-aromatic ring.

2.4.   Drawing Conventions for Organic Molecules

Several different conventions for drawing organic molecules are in force today. Consider the following drawings of butane.

1. A full structural drawing with all atoms and bonds.

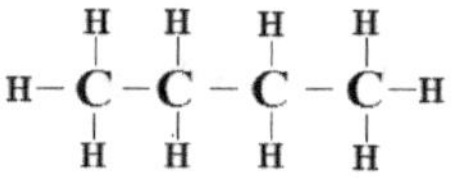

2. Full structural drawing with the hydrogen molecules suppressed.

$$C-C-C-C$$

3. A hydrogen suppressed graphical form where each vertex is a carbon and each edge is a bond.

Let's do some naming exercises!

Using the six rules previously listed and the concept of hydrogen suppressed graphical forms name the following compounds.

Problem 2.1

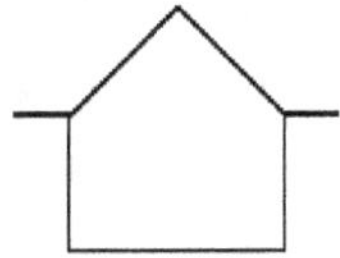

The longest chain is 5 so *pentane*. There are 2 methyl groups attached so *dimethyl*. The lowest numbering is 2,4 so *2,4-Dimethylpentane*.

Problem 2.2

The longest chain is 6 so *hexane*. There are 3 methyl groups attached so *trimethyl*. The lowest numbering is 2,3,4 so *2,3,4-Trimethylhexane*.

Problem 2.3

The molecule is cyclic so *cyclo*. There are 5 carbons so *cyclopentane*. There are 2 methyl groups attached so *dimethycyclopentane*. The lowest numbering is 1,3 so *1,3-Dimethylcyclopentane*.

Problem 2.4

The longest chain is 7 so *heptane*. There are 2 methyl groups so *dimethylheptane*.

There is an ethyl group so *ethyldimethylheptane*. The lowest numbers are 2,3,4 so *4-Ethyl-2,3-dimethylheptane*. Now let's switch from naming a hydrogen suppressed graphical representation of an alkane to drawing that same hydrogen suppressed representation if you are given a valid compound name.

Problem 2.5

Draw 2,3-Dimethylhexane.

First draw a six (hex=6) carbon chain.

Attach methyl groups to carbons 2 and 3.

Problem 2.6

Draw 4-Ethyl-3-methyl octane.

First draw an eight (oct=8) carbon chain.

Attach an ethyl group to carbon 4.

Attach a methyl group to carbon 3.

Problem 2.7

Draw 1,2-Dimethylcyclohexane. Draw a cyclohexane ring.

Attach two methyl groups adjacent to each other starting with any carbon.

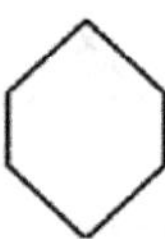

Problem 2.8

Draw 1,3-Dimethylcyclobutane. First draw 4 (but=4) carbon ring.

Attach two methyl groups at opposing corners of the ring.

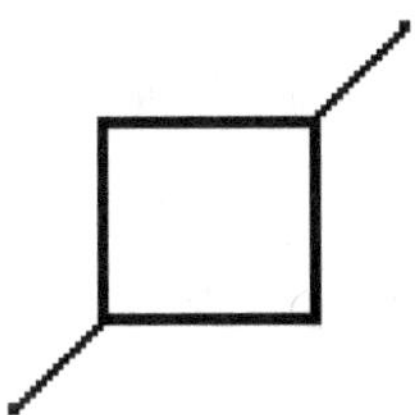

2.5.   Properties of Alkanes

Solubility–Alkanes are nonpolar and insoluble in water. They are soluble in other nonpolar solvents. (Fuel and water do not mix.)

Density–Alkanes have densities from 0.62 g/ml to 0.79 g/ml. (Fuel floats on water.)

Combustibility–All alkanes will burn readily in oxygen. (Fuel burns.)

2.6.   Combustion of Alkanes

When alkanes undergo complete combustion in oxygen they do so according to the following formula:

$$\text{Alkane} + O_2 \rightarrow CO_2 + H_2O + \text{Heat}$$

Write and balance the combustion reaction for propane.

1) We know propane has 3 carbons and $(2 \times 3) + 2 = 8$ hydrogens.

2) So the combustion reaction $C_3H_8 + O_2 \rightarrow CO_2 + H_2O$

3) Balance C products to reactants. $C_3H_8 + O_2 \rightarrow 3\,CO_2 + H_2O$

4) Balance H products to reactants. $C_3H_8 + O_2 \rightarrow 3\,CO_2 + 4\,H_2O$

5) Balance O reactants to products. $C_3H_8 + 5\,O_2 \rightarrow 3\,CO_2 + H_2O$

Let's do some combustion reaction exercises!

Problem 2.9

Write and balance the combustion reaction for 2,3 dimethylheptane.

heptane = 7 2 × methyl= 29 C's and (2 × 9) + 2 = 20 H's

$$C_9H_{20} + 14O_2 \rightarrow 9CO_2 + 10H_2O$$

Problem 2.10

Write and balance the combustion equation for the following compound.

7 C's and (2 × 7) + 2 = 16 H's

$$C_7H_{16} + 11O_2 \rightarrow 7CO_2 + 8H_2O$$

Problem 2.11

Write and balance the combustion reaction for 4-ethyl-2-methylhexane.

Carbons = 6 (hexane) + 2 (ethyl) + 1 (methyl) for a total of 9 carbons.

Hydrogens = (2 × 9) + 2 for 20 hydrogens

$$C_9H_{20} + 14O_2 \rightarrow 9CO_2 + 10H_2O$$

2.7. The Alkenes, Alkynes, and Aromatics

2.7.1. Alkenes

Alkenes are essentially alkanes that contain one or more double bonds (also a point of unsaturation). The compound 1-hexene is a straight chain of six carbon atoms with a double bond between the first and second carbons.

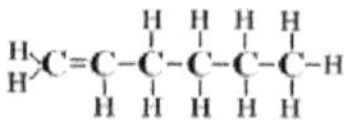

1-Hexene

C_nH_{2n}

Figure 2.5: 1-Hexene

All single bonds between carbon atoms are referred to as sigma (σ) bonds. An additional bond between two carbon atoms is referred to as a pi (π) bond. If the alkene has two double bonds it is referred as a diene.

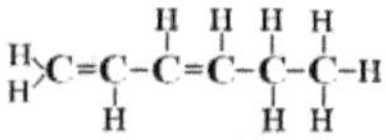

1,3-Hexadiene

C_nH_{2n-2}

Figure 2.6: 3-Hexadiene

The every other carbon π-bond configuration is known as a conjugated π-bond system. Conjugated π-bond systems are often responsible for color and odor in organic molecules.

2.7.2. *Alkynes*

Alkynes are essentially alkanes that contain one or more triple bonds between carbon atoms.

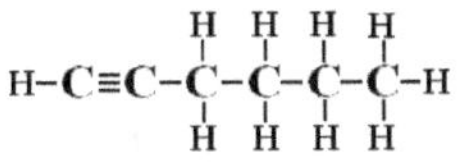

$$H-C\equiv C-\overset{\overset{\displaystyle H}{|}}{\underset{\underset{\displaystyle H}{|}}{C}}-\overset{\overset{\displaystyle H}{|}}{\underset{\underset{\displaystyle H}{|}}{C}}-\overset{\overset{\displaystyle H}{|}}{\underset{\underset{\displaystyle H}{|}}{C}}-\overset{\overset{\displaystyle H}{|}}{\underset{\underset{\displaystyle H}{|}}{C}}-H \qquad \textbf{1-Hexyne}$$

$$C_n H_{2n-2}$$

Figure 2.7: 1-Hexyne

The triple bond is a bond consisting of two σ-electrons and four π-electrons. Alkynes are essential in the makeup of the planet. A simpler way of stating that is that it takes alkynes to make a world.

2.7.3. *Nomenclature Update*

The nomenclature rules previously listed in this chapter apply to alkenes, and alkynes with the following addenda.

1.   Alkenes end with *-ene*
2.   Alkynes end with *-yne*
3.   Both use the same lowest carbon number scheme to indicate the positions of multiple bonds. (In numbering schemes double bonds take precedence over substituents.)
4.   Both use Greek prefixes to indicate more than one of the same type of multiple bond.

Problem 2.12

Name the following compound.

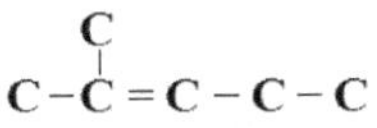

$$C-\overset{\overset{\displaystyle C}{|}}{C}=C-C-C$$

A double bond so *ene*. 5 carbons for the longest chain so *pentene*. Double Bond originates at #2 C so *2-pentene*.

A methyl group on #2 carbon so *2-methyl-2-pentene*

Problem 2.13

Name the following compound.

$$C-C=\underset{\underset{\displaystyle C}{|}}{C}-C-C$$

A double bond so *ene*. 5 carbons for longest chain so *pentene*. Double bond originates at #2 C so *2-pentene*. A methyl group on #3 C so *3-methyl-2-pentene*.

Problem 2.14

Name the following compound.

$$C-C \equiv C - \underset{\underset{\displaystyle C}{|}}{C} - C - C$$

A triple bond so *yne*. Longest chain of 6 C's so *hexyne*. Triple bond originates at #2 C so *2-hexyne*. A methyl group on #4 C so *4-methyl-2-hexyne*.

Problem 2.15

Name the following compound.

$$C=C - \underset{\underset{\displaystyle C}{|}}{C} = C \quad C$$

Two double bonds so *diene*. Longest chain of 5 C's so *pentadiene*. Double bonds originating at #1 C and #3 C so *1,3-pentadiene*.

A methyl group on # 3 C so *3-methyl-1,3-pentadiene*.

Problem 2.16

Name the following compound.

A 6 carbon cycle with a double bond so *cyclohexene*.

Double bond originates at #1 C so still *cylohexene* due to symmetry. A methyl group on #4 C so *4-methylcyclohexene*.

Problem 2.17

Name the following compound.

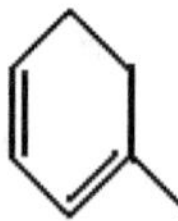

Two double bonds so diene. Double bonds originating at #1 C and #3 C so 1,3-diene. 6 C ring so cyclohexa- 1,3-diene. A methyl group so methylcyclohexa-1,3-diene. Methyl group at #1 C so 1-methylcyclohexa- 1,3-diene.

cis – trans Isomers

Unlike single bonds, double bonds are "rigid" and do not allow substituents to rotate freely about the bond. Because of this it matters which direction a substituent "approaches" the double bond.

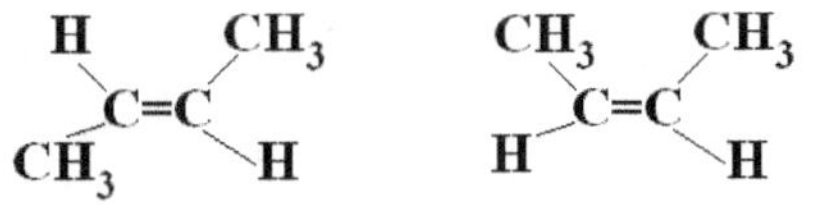

trans-2-Butene

(mp = 106 C ; bp = 0.3 C)

cis-2-Butene

(mp = 139 C ; bp = 3.7 C)

Figure 2.8: cis-trans Isomers of 2-Butene

Problem 2.18

Name the following compound.

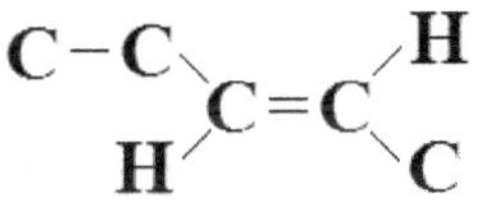

A double bond so ene. 5 carbons so pentene. Double bond originates from #2 C (from the right) so 2-pentene. Hydrogens are opposite each other so trans-2-pentene.

Problem 2.19

Name the following compound.

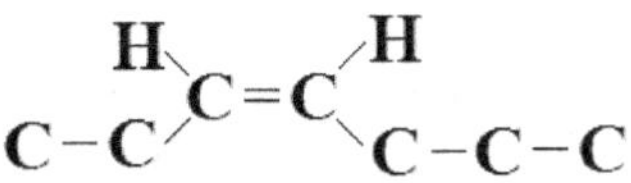

A double bond so ene. 7 carbons so heptene. Double bond originates from #3 C (from the left) so 3-heptene. Hydrogens are on the same side so cis-3-heptene. Addition Reactions with Double Bonds.

There are three common types of addition reactions that occur with or "across" double bonds. In this case the double bond is taken away.

1. Hydrogenation – the addition of hydrogen
2. Hydration – the addition of water
3. Polymerization – the addition of a monomer unit

2.8.   Hydrogenation

When hydrogen (H_2) is added across a double bond it acts to saturate the double bond by replacing it with two H's.

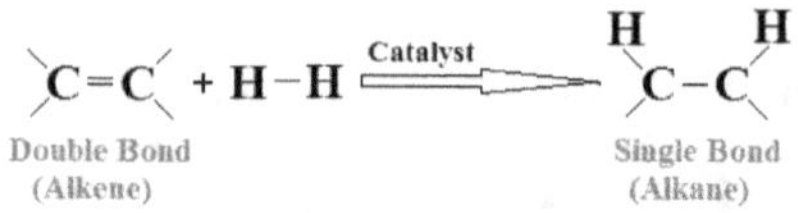

$$C=C \ + H-H \xrightarrow{\text{Catalyst}} C-C$$

Double Bond
(Alkene)

Single Bond
(Alkane)

Figure 2.9: Hydrogenation

2.8.1. Hydration

When water is added across a double bond it acts to replace the double bond with a hydrogen (H) and an alcohol group (OH).

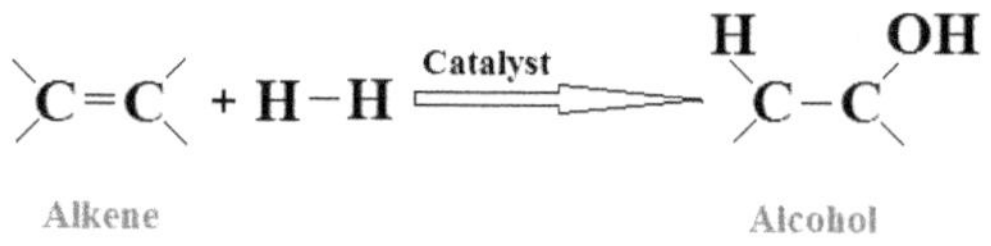

$$C=C \ + H-H \xrightarrow{\text{Catalyst}} C-C$$

Alkene

Alcohol

Figure 2.10: Hydration

2.8.2. Addition Polymers

Polymers are formed from alkenes when individual units of the alkene, known as *monomers*, add across the double bond of another monomer.

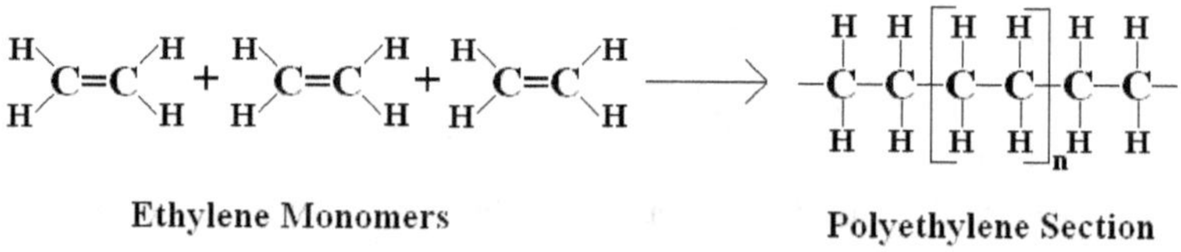

Ethylene Monomers

Polyethylene Section

Figure 2.11: Addition Polymers

Where "n" is the number of times the unit of the polymer repeats.

2.9.   Aromatic Compounds

Aromatics are compounds consisting of one or more six carbon rings that contain systems of conjugated (every other carbon) π-bond electrons. In the drawing of benzene, the simplest aromatic compound, there are three double bonds and a ring for a total of four points of unsaturation.

The general formula for all single ring aromatics is C_nH_{2n-6m} where "n" is the number of carbons and "m" is the number of aromatic rings.

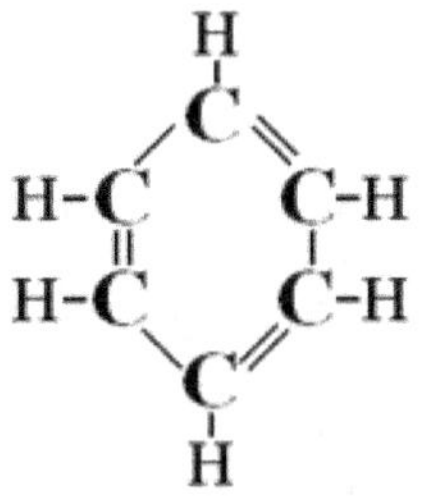

Figure 2.12: Benzene

2.9.1. The True Aromatic Ring Structure

The structure shown in Figure 11.2.8, for benzene, was envisioned by the German organic chemist Friedrich August Kekule von Stradonitz in the late 1800s.

With modern analytical techniques, capable of mapping electrostatic charges, it was determined that the Kekule structure was not entirely correct.

The extra electrons shown in the double bonds of the Kekule structure are actually π electrons but they exist as a *cloud* of π electrons around the benzene ring, which is represented as a circle in the center of the six carbon ring.

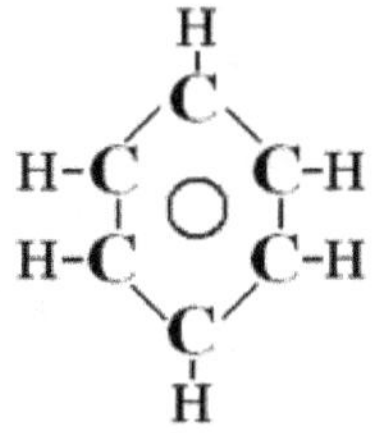

Figure 2.13: Benzene's True Structure

2.9.2. Aromatic Nomenclature

I prefer to use the line drawing method for ring structures but not for the substituents attached to the ring. For example, the second drawing below is seen more often for the molecule methylbenzene (aka, toluene).

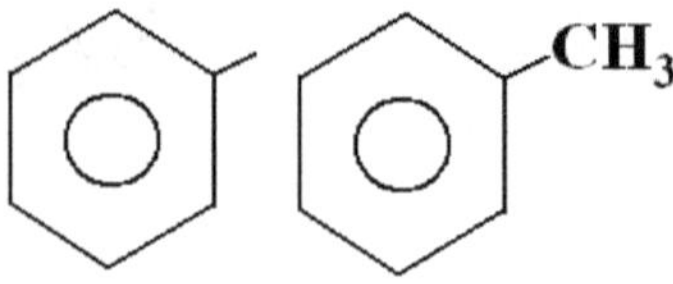

Figure 2.14: Benzene Drawings

Naming Aromatic Compounds

1. Single ring aromatic compounds with a single substituent are named as benzene derivatives. However, trivial names generally dominate.

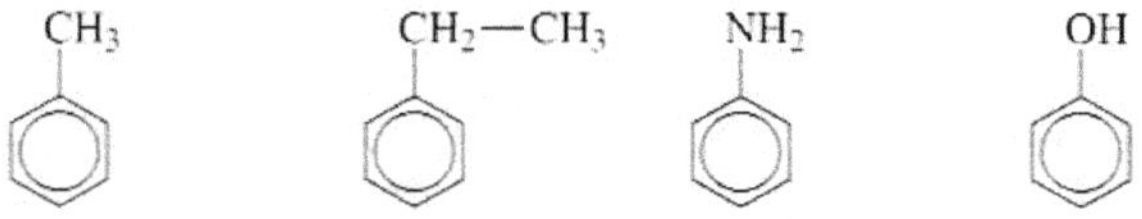

Tolulene (methylbenzene) **Ehylbenezene** **Aniline (benzenamine)** **Phenol (hydroxybenzene)**

Figure 2.15: Benzene Derivatives

2. When benzene is used as a substituent (C_6H_5 -), it is named as a phenyl group.

Phenyl group

3-Phenyl-1-butene

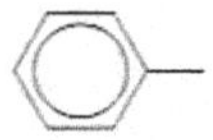

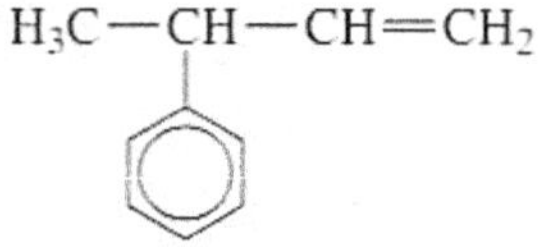

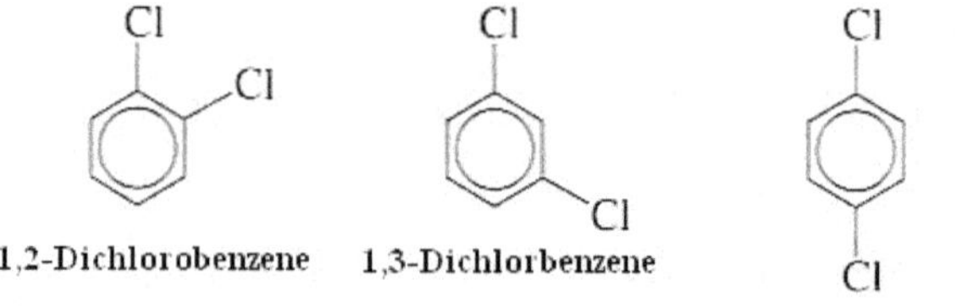

Figure 2.16: Benzene Substituents

3. When two or more substituents are present, they are numbered with the lowest numbers.

1,2-Dichlorobenzene

1,3-Dichlorbenzene

1,4-Dichlorobenzene

Figure 2.17: Benzene Substituent Numbering

4. When trivial names are used, the carbon attached to the group inherent to the trivial name is numbered as carbon 1.

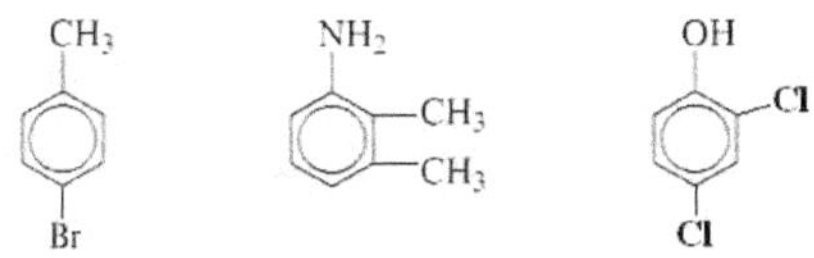

4-Bromotoluene   **2,3-Dimethylaniline**   **2,4-Dichlorophenol**

Figure 2.18: Trivial Benzene Substituent Names

5. Substituents to a benzene ring are named alphabetically.

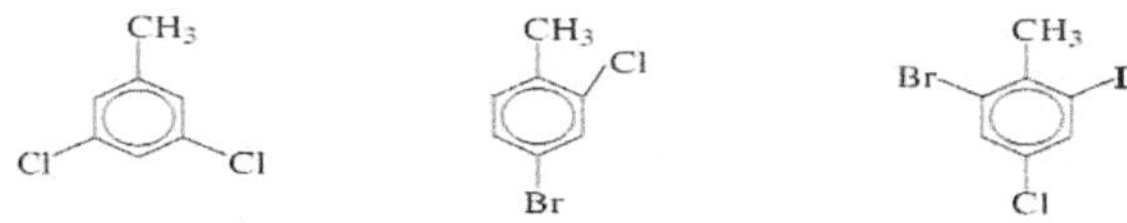

3,5-Dichlorotoluene 4-Bromo-2-chlorotoluene 2-Bromo-4-chloro-6-iodotoluene

Figure 2.19: Benzene Substituent Priorities

Problem 2.20

Name the following aromatic compound as a derivative of toluene.

The benzene ring carbon with the CH3 group is #1 C because the compound is named as a toluene. So *chlorotoluene*.

Problem 2.21

Name the following aromatic compound as a derivative of phenol.

The benzene ring carbon with the OH group is #1 C. So *3-bromophenol*

Problem 2.22

Name the following aromatic compound as a derivative of benzene.

Number the bottom Br as 1 and the top Br as 2 so that the methyl group has a lowest number. So *1,2-dibromo-4-methylbenzene.*

Problem 2.23

Name the following aromatic compound as a derivative of phenol.

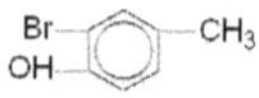

The carbon with the OH attached is #1 C. Br is #2 C and CH3 is #4 C. So *2-bromo-4-methylphenol*.

2.10. Functional Groups: Alcohols, Ethers, Aldehydes, and Ketones

Organic Functional Groups

The notion of functional groups in organic molecules and the notion of heteroatoms in organic molecules are one in the same.

All Functional Groups Contain at Least One Heteroatom

Oxygen plays a prominent role in organic functional groups. In fact, seven of eight functional groups you will be responsible for contain oxygen.

Actually, these eight oxygen containing functional groups could be looked at as differing degrees of oxidation of hydrocarbons

Alcohols

Alcohols are hydrocarbons where a hydrogen has apparently been replaced with an *-OH* group.

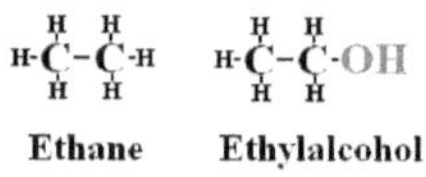

Ethane   Ethylalcohol

Figure 2.20: Alcohol

An alcohol could also be viewed as an organic derivative of water where one hydrogen has been replaced with an organic substituent.

Naming Alcohols

1. For alkane alcohols, replace the -e in the name of the longest chain containing the -OH group with *-ol*.
2. Number the chain starting at the end closest to the -OH group.
3. Name and number all other substituents relative to the -OH group.
4. For cyclic alkane alcohols, replace the -e in the cyclic alkane name with *-ol*.
5. The IUPAC name for the alcohol of benzene is *phenol*.

Problem 2.24

Name this alcohol compound

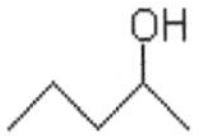

2-Pentanol

Problem 2.25

Name this alcohol compound.

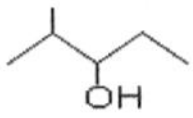

2-Methyl-3-pentanol

Problem 2.26

Name this alcohol compound.

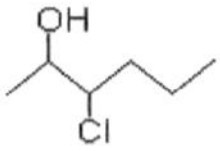

2-Chloro-2-hexanol

Problem 2.27

Draw the alcohol compound cylcopentanol.

Problem 2.28

Draw the alcohol compound 3,4-dimethylphenol.

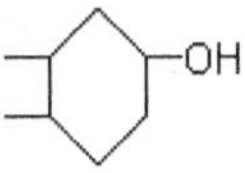

Problem 2.29

Draw the alcohol compound 2,5-dichlorophenol.

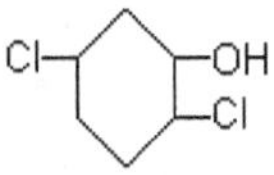

2.11. Classification of Alcohols

Alcohols are classified as primary (1°), secondary (2°), or tertiary (3°) depending on the number of carbon groups attached to the carbon that is attached to the alcohol group.

$$
\begin{array}{ccc}
CH_3 & CH_3 & CH_3 \\
H\text{-}C\text{-}H & H\text{-}C\text{-}CH_3 & H_3C\text{-}C\text{-}CH_3 \\
OH & OH & OH \\
\text{Primary} & \text{Secondary} & \text{Tertiary} \\
\text{Alcohol} & \text{Alcohol} & \text{Alcohol}
\end{array}
$$

Figure 2.21: Classification of Alcohols

Ethers

Ethers can be viewed as derivatives of water where both hydrogens have been replaced with organic substituents.

$$
\begin{array}{cccc}
H & H & H & H \\
H\text{-}C\text{-} & C\text{-}O\text{-} & C\text{-} & C\text{-}H \\
H & H & H & H
\end{array}
$$

Diethylether

Figure 2.22: Ethers

The most common ether, diethylether, was used as an anesthetic gas until safer, nonflammable alternatives weredeveloped.

Naming Ethers

Since the oxygen of an ether molecule cannot be terminal on the molecule, the rest of the molecule is named as constituents (in alphabetical order) to the ether oxygen. Look at the diethylether above. When ethers become more complicated the ether portion of the name is dropped. Below are some ether molecules used as anesthetics.

Forane   **Ethrane**   **Penthrane**

Figure 2.23: Trivial Names of Ethers

Properties of Alcohols and Ethers

Table 1 shows some properties of an alcohol and an ether of identical molar mass. These properties are compared to a pure hydrocarbon of similar mass.

Table 1: Properties of an Alcohol and Ether

Compound	Empirical Formula	Molar Mass (g/mol)	Boiling Point (°C)	Solubility (g/L)
1-Butanol	C4H10O	74	118	74
Diethylether	C4H10O	74	35	69
n-Pentane	C5H12	72	36	Insol.

Given the boiling point of the n-pentane of 36°C, ask these two questions:

Why does the n-butanol have a much higher boiling point than the n-pentane? The answer is it's ability to hydrogen bond.

Why does the diethylether have a boiling point nearly identical to the n-pentane? The answer is it's inability to hydrogen bond.

Of course the pentane would be insoluble in water. Oil and water don't mix.

Reactions of Alcohols

Figure 2.24 shows the dehydration of an alcohol to form an alkene.

H-C-C-C-H $\xrightarrow{\;H^+\;\triangle\;}$ H-C-C-C=C + H_2O

n-Propanol   **n-Pentene**   **Water**

Figure 2.24: Dehydration of Alcohol

The triangle underneath the arrow represents the use of heat.

In organic chemistry water is know as a very good *leaving group*.

Oxidation of Alcohols

An alcohol is formed by the chemical oxidation, with the symbol [O], of an alkane. This reaction is not to be confused with combustion which is also an oxidative process.

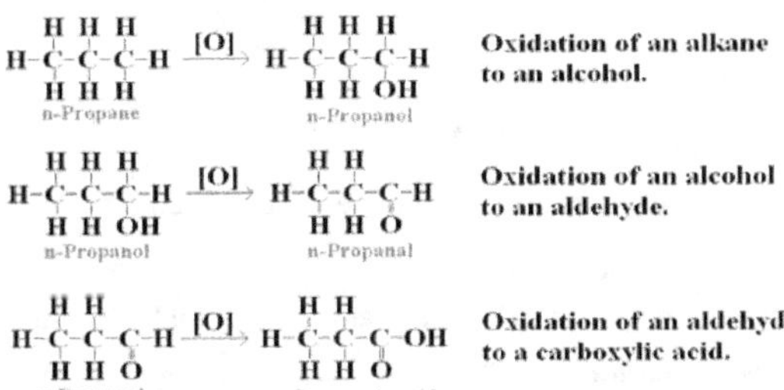

Figure 2.25: Chemical Oxidation of Alcohol

Please note that in the oxidation equations above that there is one carbon to oxygen bond in the alcohol, two carbon to oxygen bonds in the aldehyde, and three carbon to oxygen bonds in the carboxylic acid.

Not every concept introduced in this text makes sense right away. But it should be intuitive that the more times a carbon atom is bonded to an **oxygen** the more **oxidized** that atom (and hence the molecule) becomes.

Aldehydes

Aldehydes are characterized by a carbonyl group with *at least* one hydrogen atom attached to the carbonyl carbon. If the compound in Figure 2.26 had two hydrogens so attached, it would be formaldehyde.

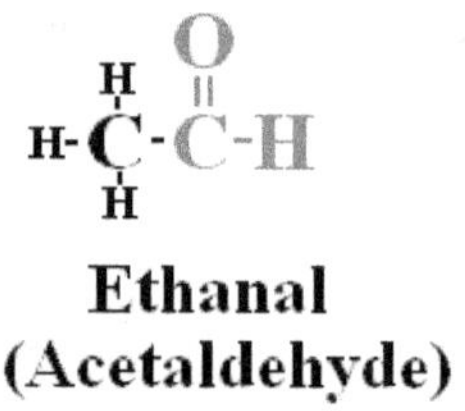

Ethanal
(Acetaldehyde)

Figure 2.26: Ethanal

Naming Aldehydes

1. Name the longest carbon chain containing the carbonyl group by replacing the -e with -*al*.

    IUPAC specifies that the aldehyde of benzene is named **benzaldehyde**.

Benzaldehyde

2. Name and number any substituents on the carbon chain by counting the carbonyl carbon as carbon #1.

Some common or trivial names of aldehydes persist. The aldehyde of methane, methanal, is more commonly named **formaldehyde**. The aldehyde of ethane, **ethanal**, is more commonly named **acetaldehyde**.

The following exercises are for naming and drawing aldehydes.

Problem 2.30

Name this aldehyde compound.

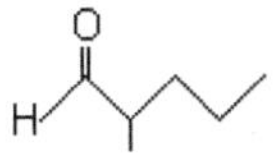

2-Methylpentenal

Problem 2.31

Name this aldehyde compound.

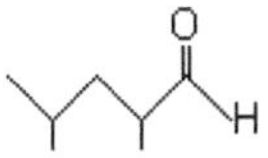

2,4-Dimethylpentenal

Problem 2.32

Draw the aldehyde compound *3-Chloro-2,4-dimethylpentenal.*

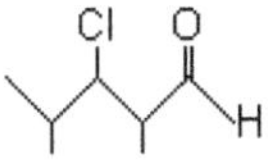

Problem 2.33

Draw the aldehyde compound *2-Chlorobenzaldehyde.*

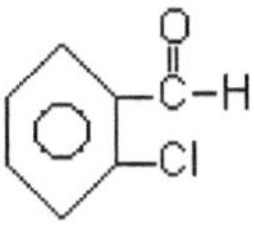

Ketones

Ketones are characterized by two carbon atoms bonded to a carbonyl group. Ketones, like aldehydes, are very often referred to by trivial names.

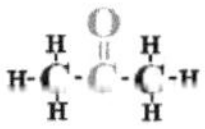

Propanone
(Acetone)

Figure 2.27: Propanone

Naming Ketones

1. Name the longest carbon chain containing the carbonyl group by replacing the -e with -*one*.
2. Number the main chain by starting from the end nearest to the carbonyl group,
3. Name and number any substituents on the carbon chain.
4. Name the carbon chain opposite the main chain as a   substituent.
5. For cyclic ketones the term *cyclo* is used in front of the ketone name.

The following exercises are for naming and drawing ketones.

Problem 2.34

Name this ketone compound.

Methylbutylketone

Problem 2.35

Name this ketone compound.

2-Methylcyclopentanone

Problem 2.36

Draw the ketone compound 4-bromo-2-methylcyclohexanone.

Problem 2.37

Draw the ketone compound 2-bromo-4-methylcyclohexanone.

Properties of Aldehydes and Ketones

The *boiling point* of aldehydes and ketones and the other functional compounds we have studied depends largely on the ability of their molecules to undergo *dipole-dipole* interactions with one another.

$$\overset{\delta+}{C}=\overset{\delta-}{O} \quad \longleftrightarrow \quad \overset{\delta+}{C}=\overset{\delta-}{O}$$

Figure 2.28: Dipole-Dipole Interactions

Remember that hydrogen bonding is just one type of dipole-dipole interaction. The figure above shows a dipole-dipole interaction between the carbonyl oxygen of one molecule to the carbonyl carbon of another of the same type molecule.

The following table, which is an expansion of Table 2 lists five hydrocarbon compounds of similar molar mass by their ascending values of boiling point.

Table 2: Dipole-Dipole Effects on Boiling Point

Compound Name	Compound Family	Compound Formula	Molar Mass (g/mol)	Boiling Point (°C)
Diethylether	Ether	C4H10O	74	35
n-Pentane	Alkane	C5H12	72	36
1-Butanal	Aldehyde	C4H8O	72	75
1-Butanol	Alcohol	C4H10O	74	138
Propanoic Acid	Carboxylic Acid	C3H6O2	74	141

Listed this way two things should become apparent.

First, the first two compounds are not capable of dipole-dipole interactions. They show similarly low boiling points for similar molar masses.

Second, the remainder of the compounds show increasing boiling points as a result of ever increasing capability for dipole-dipole interactions.

2.12. Functional Groups: Carboxylic Acids, Esters, Amines, and Amides

Carboxylic Acids

Carboxylic acids are characterized by a carbon atom and an -OH group attached to the carbonyl carbon. The hydrogen on the -OH group is known as an *acidic hydrogen*.

$$H-\overset{\overset{\textstyle H}{|}}{\underset{\underset{\textstyle H}{|}}{C}}-\overset{\overset{\textstyle O}{\|}}{C}-OH$$

Ethanoic acid

(Acetic acid)

Figure 2.29: Ethanoic Acid

Naming Carboxylic Acids

1. Identify the longest carbon chain containing the carbonyl group, and replace the -e of the alkane name with **-oic acid**.
2. Number the carbon chain, beginning with the carboxyl carbon as carbon #1.
3. Number and name any substituents on the carbon chain in reference to the #1 carbon.
4. IUPAC specifies that the carboxylic acid of benzene is named **benzoic acid**.
5. Start numbering substituents on any ringed carboxylic acid at the carbon attached to the carbonyl carbon.

The following exercises are for naming carboxylic acids.

Problem 2.38

Name this carboxylic acid.

Butanoic acid

Problem 2.39

Name this carboxylic acid.

2-Methylbenzoic acid

Problem 2.40

Name this carboxylic acid.

2-Butenoic acid

Problem 2.41

Name this carboxylic acid.

3,4-Dimethylcyclohexanoic acid

The following exercises are for the drawing of carboxylic acids.

Problem 2.42

Draw the molecule 2,4-dimethylheptanoic acid.

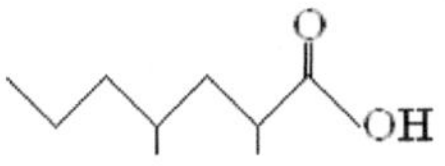

Problem 2.43

Draw the molecule 3-bromobenzoic acid.

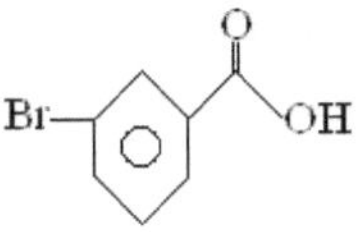

Problem 2.44

Draw the molecule 2,4-hexadienoic acid.

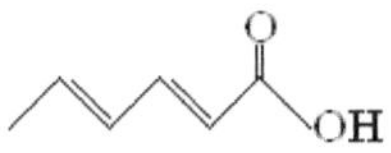

Problem 2.45

Draw the molecule 3-bromo-2-chlorocyclopentanoic acid.

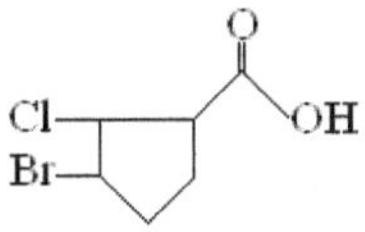

Carboxylic acids are among the most polar of hydrocarbons since they contain two polar groups in each molecule. Consider the structure of propanoic acid.

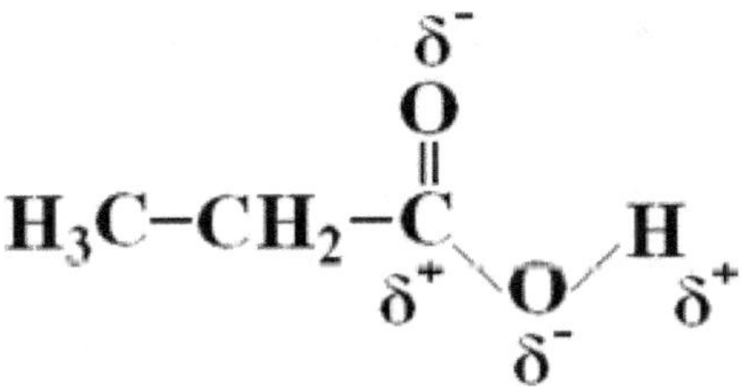

Figure 2.30: Propanoic Acid

Because of the two polar groups, it is possible for one acid molecule to bond to another. Consider the dimer of two ethanoic acid molecules

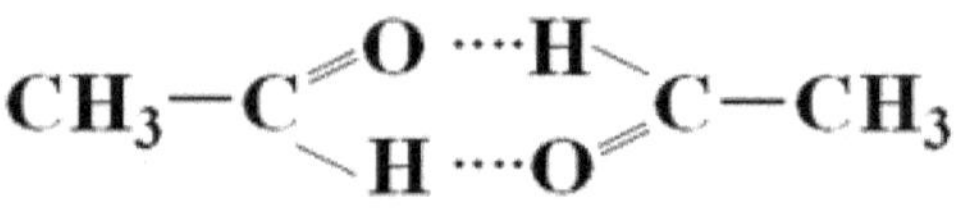

$$CH_3-C{\overset{O\cdots H}{\underset{H\cdots O}{}}}C-CH_3$$

Figure 2.31: Ethanoic Acid Dimer

Properties of Carboxylic Acids

Table 3 which includes a carboxylic acid, shows the effect of dipole-dipole interactions on the boiling points of compounds with similar molar masses.

Reactions of Carboxylic Acids

The ***esterification*** of a carboxylic acid with an alcohol is a ***condensation*** reaction.

$$CH_3CH_2\overset{O}{\overset{\|}{C}}OH + CH_3CH_2CH_2OH \underset{}{\overset{H^+}{\rightleftharpoons}} CH_3CH_2\overset{O}{\overset{\|}{C}}OCH_2CH_2CH_3 + H_2O$$

$$\text{Propanoic Acid} \qquad \text{Propanol} \qquad\qquad \text{Propylpropanoate} \qquad \text{Water}$$

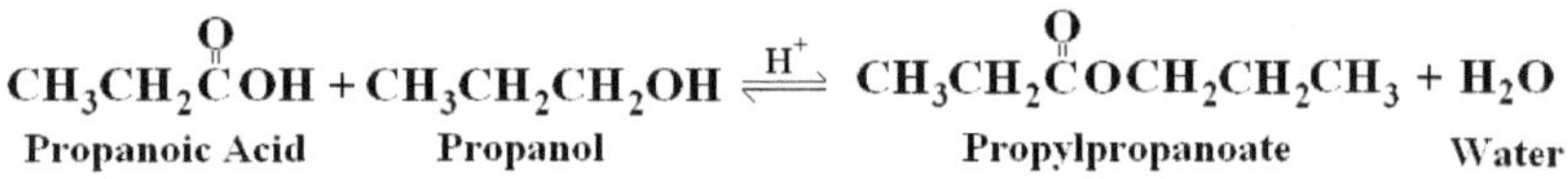

Figure 2.32: Esterification of Propanoic Acid

Notice the occurrence of water in the products of the reaction. Remember, this is a ***condensation*** reaction. The ***neutralization*** of a carboxylic acid occurs when the acid is reacted with a basic compound.

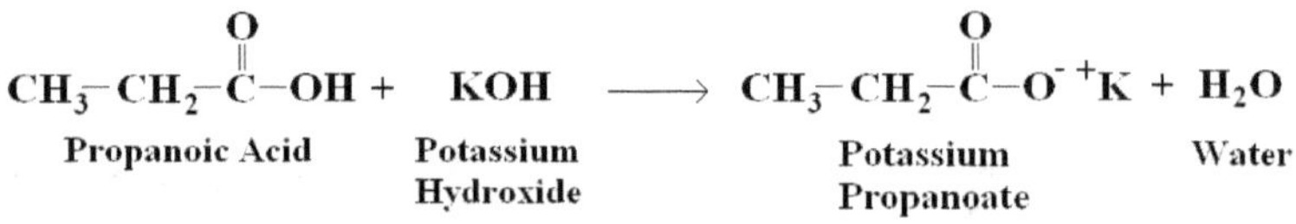

$$CH_3-CH_2-\overset{O}{\overset{\|}{C}}-OH + KOH \longrightarrow CH_3-CH_2-\overset{O}{\overset{\|}{C}}-O^-\,{}^+K + H_2O$$

$$\text{Propanoic Acid} \qquad \text{Potassium Hydroxide} \qquad \text{Potassium Propanoate} \qquad \text{Water}$$

Figure 2.33: Neutralization of Propanoic Acid

The ***ionization*** of a carboxylic acid in an aqueous solution is the manner in which the compound acts as an acid.

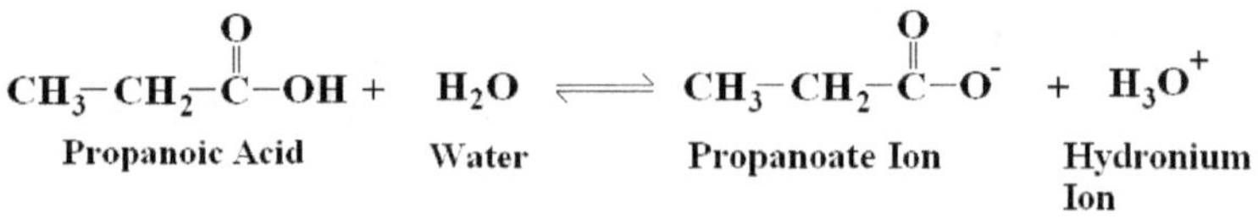

$$CH_3-CH_2-\overset{O}{\overset{\|}{C}}-OH + H_2O \rightleftharpoons CH_3-CH_2-\overset{O}{\overset{\|}{C}}-O^- + H_3O^+$$

$$\text{Propanoic Acid} \qquad \text{Water} \qquad \text{Propanoate Ion} \qquad \text{Hydronium Ion}$$

Figure 2.34: Ionization of Propanoic Acid

Esters

Esters are characterized by a carbon atom group replacing the acidic hydrogen of the corresponding carboxylic acid.

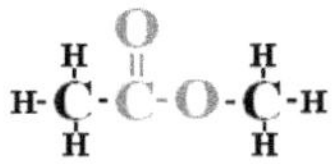

Methylethanoate
(Methlyacetate)

Figure 2.35: Methylethanoate

Naming Esters

1. Name the carbonyl side of the ester by replacing -oic acid with **-oate**.
2. Name the oxygen side of the ester as the alkane substituent.
3. IUPAC specifies the name of the ester of benzene as **benzoate**.
4. Start numbering on any ringed esters at the carbon attached to the carbonyl carbon.

The following exercises deal with naming and drawing ester compounds.

Problem 2.46

Name this ester compound.

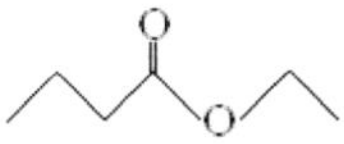

Ethylbutanoate

Problem 2.47

Name this ester compound.

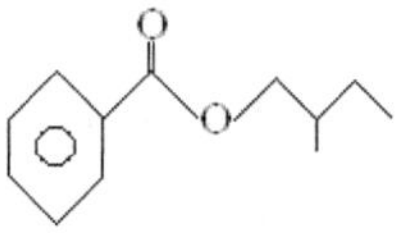

2-Methylbutylbenzoate

Problem 2.48

Draw Ethyl-2-hexenoate.

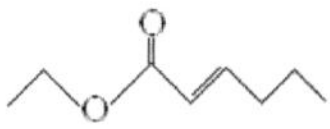

Problem 2.49

Draw 2-Propenylbenzoate.

In the chemistry of plants, esters play a major role in generating the characteristic odors and flavors of the plants.

Table 3: Ester Odors and Flavors

Odor/Flavor	Common Name	IUPAC Name
Banana	*iso*-Amylacetate	3-Methylbutylethanoate
Wintergreen	Methylsalicylate	Methyl-2-hydroxybenzoate
Pineapple	Ethylbutyrate	Ethylbutanoate
Cherry	Benzylbutyrate	Benzylbutanoate
Peach	Benzylacetate	Benzylethanoate
Apple	Methylbutyrate	Methylbutanoate
Pear	n-Propylacetate	n-Propylethanoate

Table 4 compares the boiling point and water solubility of an ester, alcohol, and two acids all near the same molecular weight.

Table 4: Comparison of Ester Properties

Compound	Molar Mass (g/mol)	Boiling Point (°C)	Aqueous Solubility
Methylethanoate	74	57	moderate
1-Butanol	74	118	complete
Propanoic Acid	74	141	complete
Ethanoic Acid	60	118	complete

Ethanoic acid was included in this table to compare only to methylethanoate.

If the acidic hydrogen of a carboxylic acid is replaced with a methyl group through esterification, the boiling point goes down while the molar mass goes up. Hydrogen bonding is no longer possible.

Once again, the effect of the dipole-dipole interaction of hydrogen bonding on boiling points and solubilities is reaffirmed. The **base hydrolysis** of an ester is analogous to the neutralization of a carboxylic acid (both reactions for a metal ion ester). The base hydrolysis forms an alcohol rather than a water as in the neutralization of a carboxylic acid.

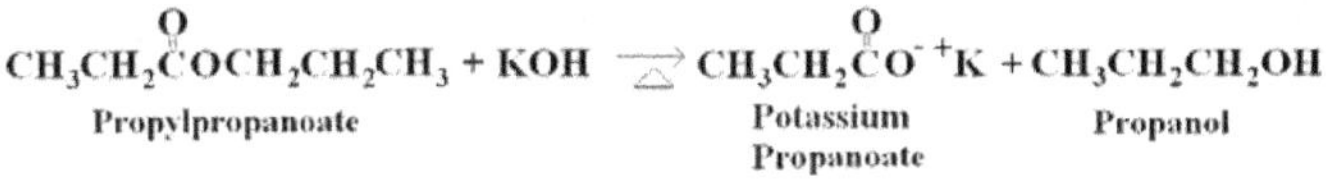

$$CH_3CH_2\overset{O}{C}OCH_2CH_2CH_3 + KOH \;\rightleftharpoons\; CH_3CH_2\overset{O}{C}O^-\,{}^+K + CH_3CH_2CH_2OH$$

Propylpropanoate   Potassium Propanoate   Propanol

Figure 2.36: Base Hydrolysis of an Ester

Amines

Amines are organic derivatives of ammonia in the same way as alcohols and ethers are organic derivatives of water.

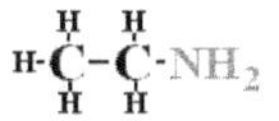

$$H\text{-}\overset{\overset{H}{|}}{\underset{\underset{H}{|}}{C}}\text{-}\overset{\overset{H}{|}}{\underset{\underset{H}{|}}{C}}\text{-}NH_2$$

Ethylamine

Figure 2.37

One distinction of amines is that whether one or all of the ammoniacal hydrogens are replaced with organic substituents the compound is still an amine.

Amines are characterized by the number of carbon atoms attached to the nitrogen atom as follows.

One carbon atom attached makes a ***primary amine***. Two carbon atoms attached make a ***secondary amine***. Three carbon atoms attached makes a ***tertiary amine***.

CH_3NH_2 Methylamine (A Primary Amine)

CH_3NHCH_3 Dimethylamine (A Secondary Amine)

CH_3NCH_3 Trimethylamine (A Tertiary Amine)
$\quad\ |$
$\quad CH_3$

Figure 2.38: Primary (1°), Secondary (2°), and Tertiary (3°) Amines

Naming Amines

1. Everything attached to the nitrogen is named as a substituent.
2. A carbon connected to a nitrogen is carbon #1. Name and number substituents "to a substituent" in reference to the #1 carbon.
3. The prefixes ***di*** and ***tri*** are used to identify two or three identical groups attached to the nitrogen.
4. IUPAC specifies that the amine of benzene is named ***aniline***.

Substituents to the nitrogen of an amine are prefaced with **N**. The following exercises are for naming amines.

Problem 2.50

Name this amine compound.

Butylamine

Problem 2.51

Name this amine compound.

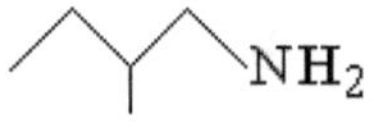

2-Methylbutylamine

Problem 2.52

Name this amine compound.

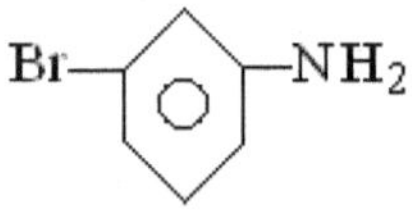

2-Bromoaniline

Problem 2.53

Name this amine compound.

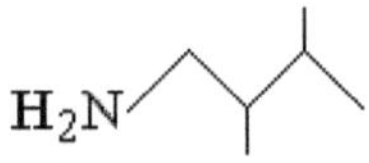

2,3-Dimethylbutylamine

Properties of Amines

Amines have **higher boiling points** than hydrocarbons of similar mass since the polar N-H bond facilitates hydrogen bonding.

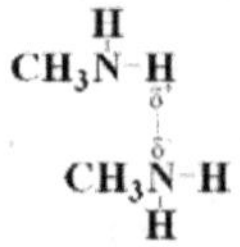

Figure 2.39: Hydrogen Bonding in Amines

Of course hydrogen bonding cannot occur in tertiary amines since there are no N-H bonds.

There is another consideration, besides hydrogen bonding, that effects the boiling points of organic molecules.

When organic molecules lay next to one another, it is less likely that intermolecular forces will hold them together the more they are branched. This effect is known as **steric hindrance**.

Observe the boiling points of the three amines in Figure 2.40. Note that each of these amines has the empirical formula $C_4H_{11}N$.

$$CH_3CH_2CH_2CH_2NH_2$$

n-Butylamine (1°) bp = 78°C

$$CH_3CH_2CH_2NHCH_3$$

N-Methylpropylamine (2°) bp = 62°C

$$CH_3CH_2NCH_3$$
$$CH_3$$

N,N-Dimetyhylethylamine (3°) bp = 36°C

Figure 2.40: Steric Hindrance Effects on the Boiling Points of Amines

Due to the hydrogen bonding in amines they are more **water soluble** than similar weight compounds that are incapable of hydrogen bonding.

If there are more than six carbons in the amine molecule, the alky portion of the molecule will diminish the water solubility afforded by the hydrogen bonding.

Amines act as **Lewis bases** in that they donate an electron pair. In Figure 2.41 the dimethylamine donates its electron pair to form the dimethylammonium ion and a hydroxide ion.

$$CH_3CH_2\ddot{N}H_2 + HOH \rightleftharpoons CH_3CH_2NH_3^+ + {}^-OH$$

Figure 2.41: Dimethylamine as a Lewis Base

Amides

Amides are characterized by a carbon atom and an amine group attached to the carbonyl carbon.

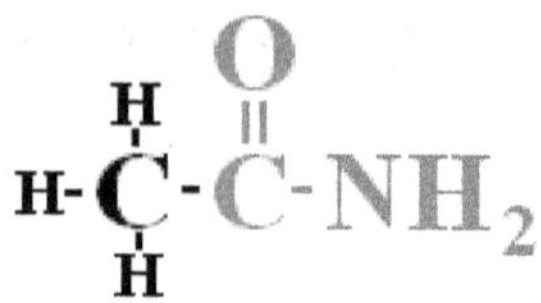

Ethanamide
(Acetamide)

Figure 2.42: Ethanamide

The same primary, secondary, and tertiary considerations of amines apply to amides.

Naming Amides

1. Drop the -ic acid or -oic acid from the carboxylic acid name and add the suffix **amide.**

2. Name and number the amide from the carboxylic carbon.

3. IUPAC specifies that the amide of benzene is named **benzamide.**

4. Name and number substituents on any ringed amide starting from the carbon connected to thecarbonylcarbon.

5. Substituents to the amide nitrogen are preceded by **N.**

Properties of Amides

Amides **do not have the basic** properties of amines. Only methanamide is liquid at room temperature.

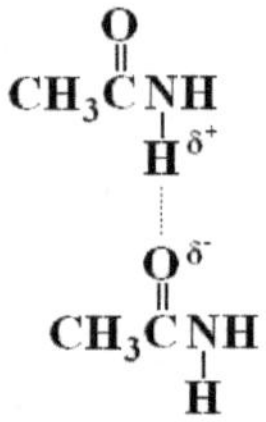

Figure 2.43: Hydrogen Bonding in Amides

Primary amides have higher **melting points** than the other amides due to the ability of an amide molecule to hydrogen bond with other amide molecules. Note that this an example of **polymerization** in a hydrocarbon.

The bond in this polymer formation is called an **amide link** by a chemist and a **peptide link** by a biochemist or biologist. The two terms are entirely synonymous.

2.13. The Concept of Aromaticity

Aromaticity is the occurrence of delocalization of what appears to be a conjugated system of π electrons in a cyclic structure. The cyclic structure will be more stable than what could be normally attributed to conjugation alone.

Aromatics

Aromatics are compounds consisting of one or more six carbon rings that contain systems of conjugated π-bond electrons.

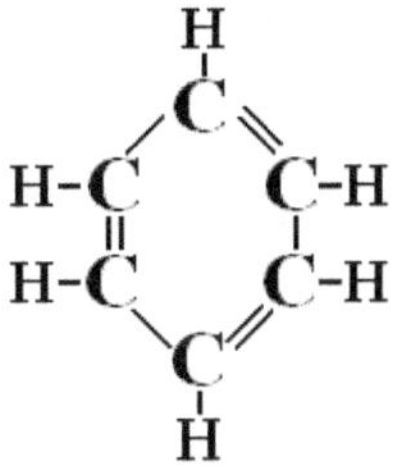

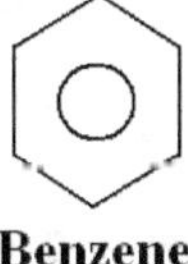

Benzene

Figure 2.44: An Aromatic Molecule

In Figure 2.44, benzene appears to have three double bonds for a total of six π electrons.

The structure shown for benzene was "envisioned" by the German organic chemist Friedrich August Kekule von Stradonitz in the late 1800s.

With modern analytical techniques, capable of mapping electrostatic charges, it was determined that the Kekule structure was not entirely correct.

The extra electrons shown in the double bonds of the Kekule structure are actually π electrons existing as a ***cloud*** of π electrons around the benzene ring.

Figure 2.45 shows a modified representation of the benzene molecule using a circle inside the ring to represent the cloud of π electrons.

Benzene

Figure 2.45: A Simplified Aromatic Ring Structure

Even though this structure is a more "honest" representation of the simplest aromatic molecule, most drawings of aromatic molecules that you will see in chemical literature still use the Kekule structure.

Aromatic Nomenclature

The benzene molecule is a perfectly symmetrical molecule. For this reason the number one carbon is not established until a substituent is added to the ring.

The IUPAC name for the six-member aromatic ring is **benzene**.

All nomenclature rules you have learned so far still apply to aromatic molecules.

The following exercises will illustrate the ease of naming and drawing aromatic compounds.

Exercise 1

Name the aromatic compound.

Methylbenzene (aka. toluene)

Exercise 2

Name the aromatic compound.

1,2-Dimethylbenzene (Note that either of the substituted carbons can be the number one carbon.)

2.14. The Concepts of Saturation and Unsaturation

Everyone has heard the term **saturated fat.** So what does the term saturation of unsaturation mean? First, one must understand what a point of unsaturation in an organic molecule actually means. Remembering the fact that the term **hydrocarbon** means a compound made up of, at least, hydrogen and carbon and that each carbon in a neutral hydrocarbon molecule must always be bonded four times.

The degree of unsaturation in an organic molecule, as compared to a completely saturated analog, then becomes an accounting of the overall amount of hydrogen attached to that molecule.

If you recall when we studied the pure hydrocarbons at the beginning of this chapter, I gave you a general formula for each of the pure hydrocarbon families.

Table 5 is a recap of those general formulae.

Table 5: General Formulae for Pure Hydrocarbons

Family Name	Compound Name	Structure	General Formula
Alkane	n-Hexane		C_nH_{2n+2}
Alkene	2-Hexene		C_nH_{2n}
Alkyne	2-Hexyne		C_nH_{2n-2}
Cycloalkane	Cyclohexane		C_nH_{2n}
Aromatic	Benzene		C_nH_{2n-6m} *

* The "m" in the general formula for aromatics signifies the number of rings in the aromatic compound.

The drawings in table6 are hydrogen suppressed graphical representations of each compound.

Using the general formulae for each compound we should be able to calculate the number of hydrogens for each compound.

The n-hexane compound has six carbons so its actual formula should be $C_6H_{(2x6)+2}$ which is C_6H_{14}.

Looking at the hydrogen suppressed graphical drawing of n-hexane and remembering the tetravalent nature of carbon it should be obvious that the two terminal carbons are CH_3 groups and the four internal carbons are CH_2 groups. This adds up to 14 hydrogens or C_6H_{14}. The utility of a general formula is that, in this case, every member of the alkane family will adhere to the C_nH_{2n+2} general formula. So the general formula for alkanes ***mathematically describes*** the entire alkane family of compounds.

The alkene and cycloalkane families each have the general formula C_nH_{2n}. Simple arithmetic reveals that alkenes and cycloalkanes each have ***two fewer hydrogens*** than their alkane analogs.

The 2-hexene compound has two fewer hydrogens than the n-hexane compound because of the presence of a double bond.

The cyclohexane compound has two fewer hydrogens than the n-hexane compound because of the presence of a ringed structure.

Questions

Part A

1. What is organic chemistry ?
2. Define hydro cabons.
3. Define Alkalies
4. List the procedure for Naming Alkanes.
5. What are the Review of Rules for Naming Alkanes ?
6. What are the properties of alkanes ?
7. Define the nomenclature updates.
8. Define Alcohols
9. What is Ethers?
10. List the procedure for naming amides.

Part B

1. Briefly explain organic familes.
2. Explain in detail about saturation and unsaturation
3. What is meant by Aromocity. Explain briefly
4. Explain the functional groups: carboxylic acids, esters, amines, and amides

CHAPTER 3

COMPLEX ORGANIC MOLECULES

3.1.   Carbohydrates: Sugars to Polysaccharides

Carbohydrates were originally thought to be a hydrated form of carbon. Early researchers were able to arrive at an empirical formula of $C_6H_{12}O_6$ for common glucose. They then interpreted this to be in the form of $C_6(H_2O)_6$. Even though the true structures of sugars and other carbohydrates are known, the term carbohydrate is still used.

Types of Carbohydrates

There are three basic types of carbohydrates:

- Monosaccharides: The basic $C_6H_{12}O_6$ unit.

- Disaccharides: Two monosaccharide units.

- Polysaccharides: Many monosaccharide units.

Monosaccharides

There are two basic monosaccharide structures.

Aldose

The carbonyl group is on the first carbon to form an aldehyde (-CHO) group.

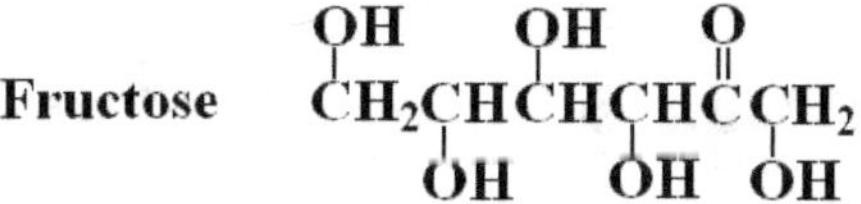

Figure 3.1: Glucose

Glucose is an ***aldose*** monosaccharide.

Ketose

The carbonyl carbon is on the second carbon to form a ketone (-CO) group.

Figure 3.2

Fructose is ***ketose*** monosaccharide.

Fischer Projections of Monosaccharides

It actually matters which side of the fifth carbon of the fructose or glucose molecule that the -OH group is attached.

If a carbon atom in an organic molecule can have four distinctly different groups attached to it, it is said to be a *chiral* atom.

Chirality is a notion of left or right handedness in a molecule. Any organic compound that has a chiral center will have left and right handed molecules known as stereoisomers.

Looking at the drawing of glucose in Figure 3.1, it appears that the fifth carbon (actually second from the right) has the following four distinct groups attached: A hydrogen, -OH, CH_2OH, and the rest of the molecule. Therefore, carbon five is chiral.

It is difficult to tell left from right in the above drawings. If you draw glucose from Figure 3.1 vertically with the aldehyde at the top, the fifth carbon will be second from the bottom.

This vertical drawing is known as a *Fischer projection*. If the -OH group on carbon number five is on the right, the stereoisomer is known as D-glucose. If the -OH group is on the left, the stereoisomer is known as L-glucose.

D and L are simply the first letters of the Latin words for right and left *dextro* and *levo*. Figure 3.3 compares the D and L stereoisomers of glucose.

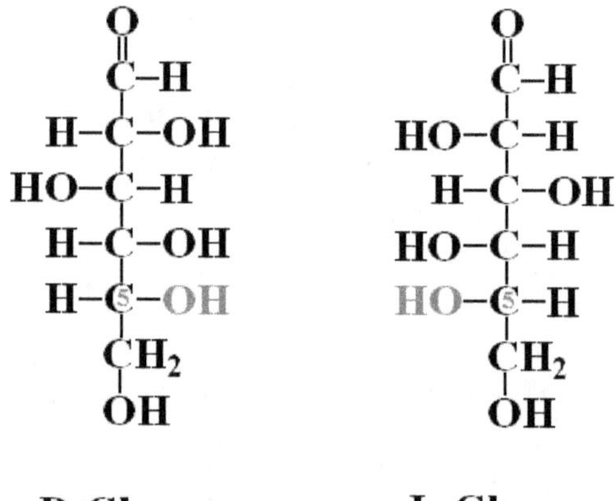

Figure 3.3: D and L Stereoisomers of Glucose

The D stereoisomers of glucose, fructose, and galactose (another monosaccharide) are more prevalent in nature and more easily used by the body.

Ring Structures of Monosaccharides

So far the monosaccharides listed have been described as open chained molecules.

The carbonyl group of a particular monosaccharide can react with an alcohol group in the same molecule to form a ring structure.

Figure 3.4 shows the comparison of the open chained structure and the ring structure of D-glucose.

**Opened Chain
D-Glucose**

**Ring Structure
D-Glucose**

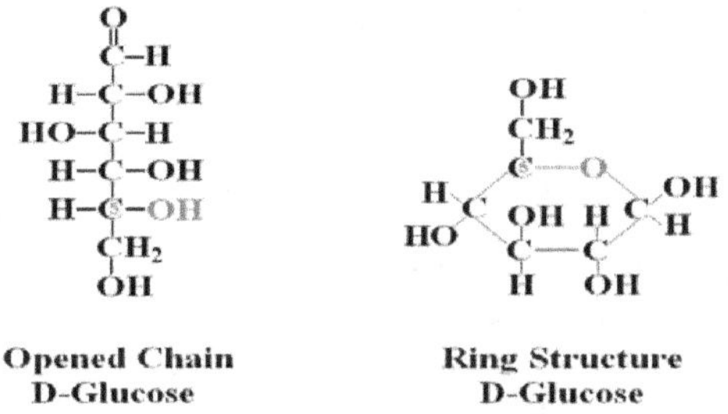

Figure 3.4: Opened Chain and Ring Structure for D-Glucose

It should be noted here that the chemical formula for the ring structure of D-glucose is still $C_6H_{12}O_6$. There is no additional point of unsaturation in the ring structure. The double bond of the carbonyl carbon in the open chained structure was sacrificed for a single ring.

Haworth Structures of Monosaccharides

The number one carbon in the ring structure of D-glucose has been rendered *chiral*. In order to distinguish whether the -OH group on that carbon is *pointing downward (α)* or *pointing upward (β)* we embrace the Haworth structures.

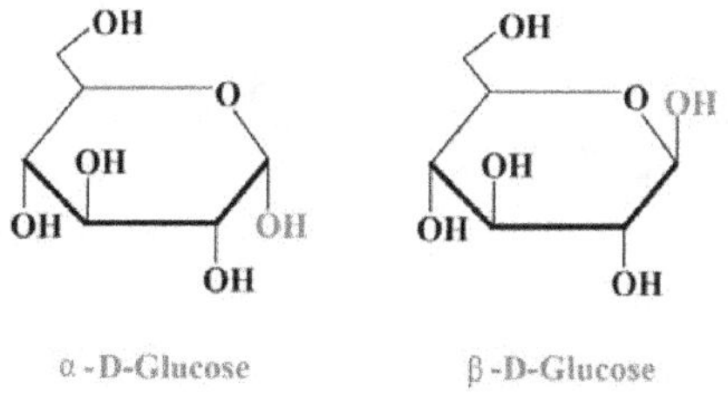

α-D-Glucose

β-D-Glucose

Figure 3.5: Hawthorn Structures of D-Glucose

The oxygen atom in the ring is still attached to the number five carbon. It is still on the right hand side of the number five carbon which establishes it as D-glucose.

Disaccharides α-Maltose

When the 1 and 4 position alcohols on two α-D-glucose molecules bond in a downward (*α*) position, α-maltose is formed. Figure 3.6 shows the formation of α-maltose from two α-D-glucose molecules.

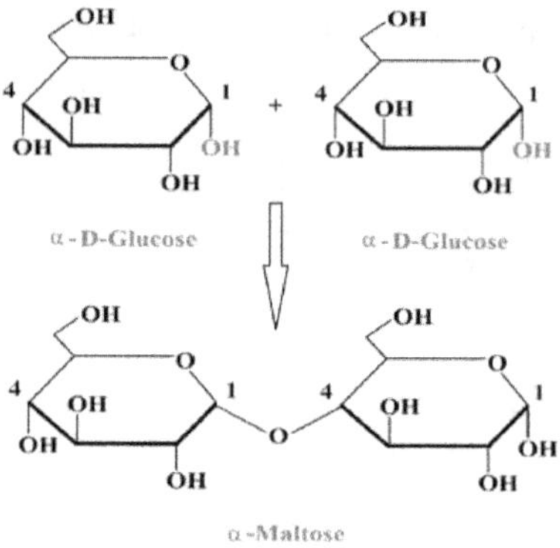

Figure 3.6: Formation of α-Maltose

The bond between the two rings is in a downward or α position between the number one carbon on one of the α-D-glucose rings and the number four carbon on the other α-D-glucose ring. Therefore, it is an *α-1,4-glycosidic bond*.

α-Lactose

When the 1 position alcohol of β-D-galactose and the 4 position alcohol of α-D-glucose bond in an upward (β) position, α-lactose is formed.

Figure 3.7 shows the formation of α-lactose.

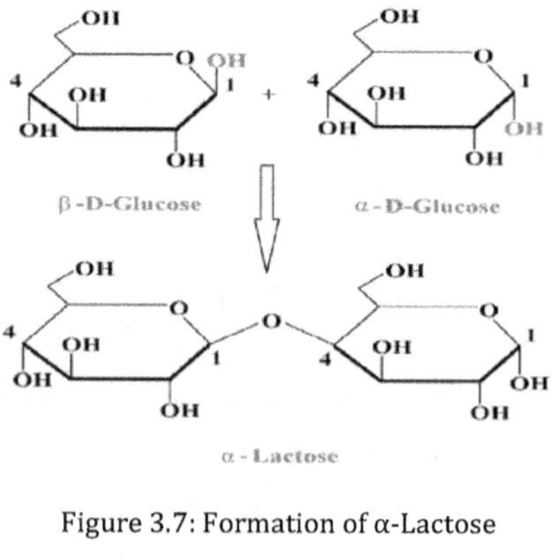

Figure 3.7: Formation of α-Lactose

It is the -OH closest to the (previously) carbonyl oxygen that determines α, β orientation. Note the **β-1,4-glycosidic bond.**

Sucrose

When the 1 position alcohol in α-D-glucose and the 2 position alcohol in β-D-fructose are joined by an α,β-glycosidic bond, sucrose is formed.

Figure 3.8 shows the formation of sucrose.

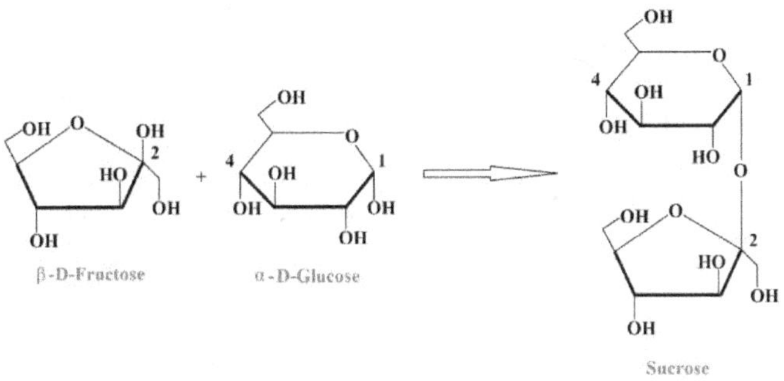

Figure 3.8: Formation of Sucrose

Note that where the oxygen connects the two monosaccharides is a *α,β-1,2-glycosidic bond.* Sucrose is a non reducing sugar since its glycosidic bond cannot open to an aldehyde.

Polysaccharides

Polysaccharides can be viewed as polymers of many monosaccharides bonded together. The position of their glycosidic bonds can result in profound differences between molecules. Four biologically important polysaccharides are *amylose, amylopectin, cellulose, and glycogen.*

Starch is composed of about 20% *amylose* and about 80% *amylopectin*. Both of these polysaccharides are characterized by *α-1,4-glycosidic* bonds. Figure 3.9 shows an unbranched chain of amylose.

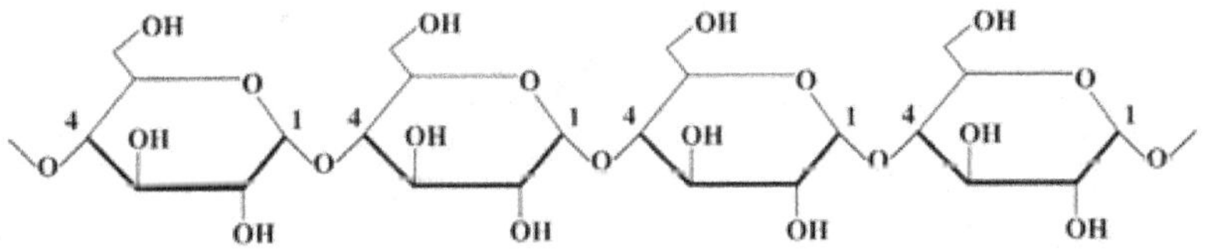

Figure 3.9: An Unbranched Chain of Amylose

Note that each pair of monosaccharides are bonded between position 1 and 4 carbons in a downward direction. This establishes the 1,4-α-glycosidic bonds.

Figure 3.10 shows a branched chain of amylpectin.

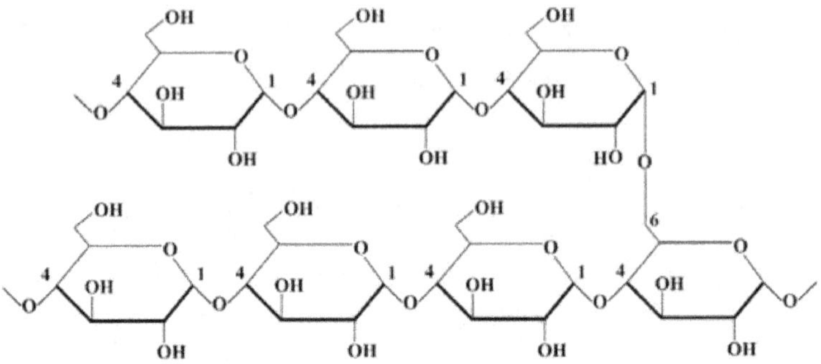

Figure 3.10: A Branched Chain of Amylopectin

Take note of the 1,6-α-glycosidic bond that connects the branched parts of the amylpectin.

3.2.   Carbohydrates: Cellulose and Glycogen

Cellulose

Cellulose is a polymer of glucose characterized by **β-1,4-glycosidic bonds**.

Cellulose is the principle structural material in plants such as the wooden or leafy portions of the plant. Cotton is nearly pure cellulose.

Figure 3.11 shows unbranched cellulose with its β -1,4-glycosidic bonds.

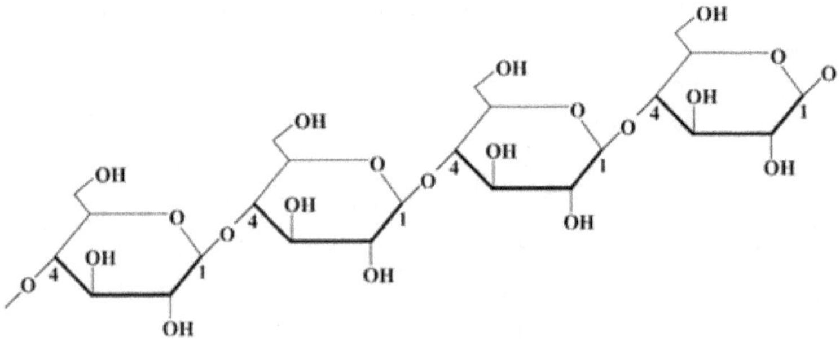

Figure 3.11: A Branched Cellulose Chain

Glycogen

Glycogen, also a polymer of glucose, is characterized by **α-1,4- glycosidic bonds**.

Glycogen, or animal starch, is stored in the muscles and liver of animals. Glycogen is hydrolyzed in the cells to help maintain blood glucose levels

Figure 3.12 shows an unbranched glycogen chain.

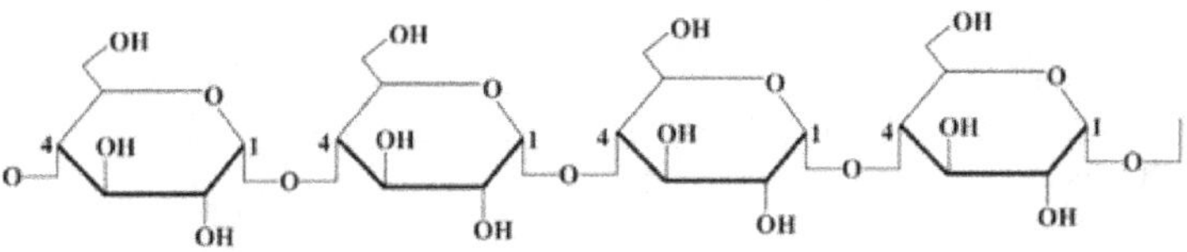

Figure 3.12: An Unbranched Glycogen Chain

3.3.   Lipids: Fatty Acids and Waxes

Lipids

Lipids are a family of biomolecules that are insoluble in water and soluble in organic solvents. The different types of lipids are characterized in the organizational chart in Figure 3.13.

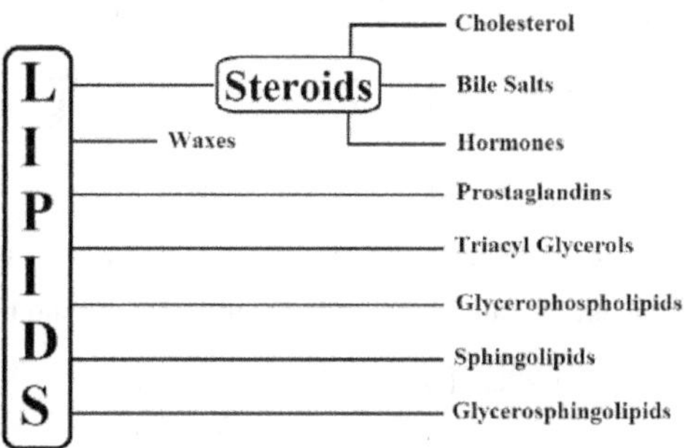

Figure 3.13: Lipids

Fatty Acids

The simplest of lipids, fatty acids, are comprised of long unbranched carbon chains terminating in a carboxylic acid groups. Table 3.1 shows some common saturated fatty acids. The trivial and IUPAC names for each fatty acid is listed. The carbon number, common source, melting point, and graphical drawing is listed for each fatty acid.

Table 3.1: Saturated Fatty Acids

Trivial Name	IUPAC Name	C#	Common Source	mp °C	Graphical Drawing
Capric Acid	Decanoic Acid	10	Coconut	32	
Lauric Acid	Dodecanoic Acid	12	Coconut	43	
Myristic Acid	Tetradecanoic Acid	14	Nutmeg	54	
Palmitic Acid	Hexadecanoic Acvid	16	Palm	63	
Stearic Acid	Octadecanoic Acid	18	Animal	70	

Monounsaturated fatty acids (MUFA) are those containing one carbon to carbon double bond. Table 3.2 shows two common monounsaturated fatty acids.

Table 3.2: Monounsaturated Fatty Acids

Trivial Name	IUPAC Name	Source	MP °C	Graphical Drawing
Palmitoleic Acid	9-Hexadeceneoic Acid	Butter	0	
Oleic Acid	9-Octadeceneoic Acid	Olive	43	

Polyunsaturated fatty acids are fatty acids with two or more carbon to carbon double bonds. Table 3.3 show polyunsaturated fatty acids of contemporary interest.

Table 3.3: Polyunsaturated Fatty Acids

Trivial Name	IUPAC Name	Type	C#	Graphical Drawing
Alpha-Linolenic Acid	*all-cis* 9,12,15 Octadecatrienoic Acid	Omega-3	18	
Gamma-Linolenic Acid	*all-cis* 6,9,12 Octadecatrienoic Acid	Omega-6	18	

The fatty acids in Table 3.3 are depicted by trivial name, IUPAC names, and their drawings. The important thing to note is the *all-cis* part of the IUPAC nomenclature.

Any fatty acid molecule is characterized by a carbonyl (acid group) end and a methylene (CH$_3$) end. The omega designation is assigned by how many carbons the nearest double bond is from the methylene carbon.

cis and trans Isomerization of Fatty Acids

When a carbon to carbon bond contains only one set of sigma electrons, the carbons can rotate freely about the bond. When there are additional pi bonds, as in the case of double bonds, the bond then becomes rigid.

If the bond is rigid it then becomes significant which way the atoms of the molecule are situated around the double bond.

In Figure 3.13 the stark difference between the shapes of *cis*-oleic acid and *trans*-oleic acid are shown.

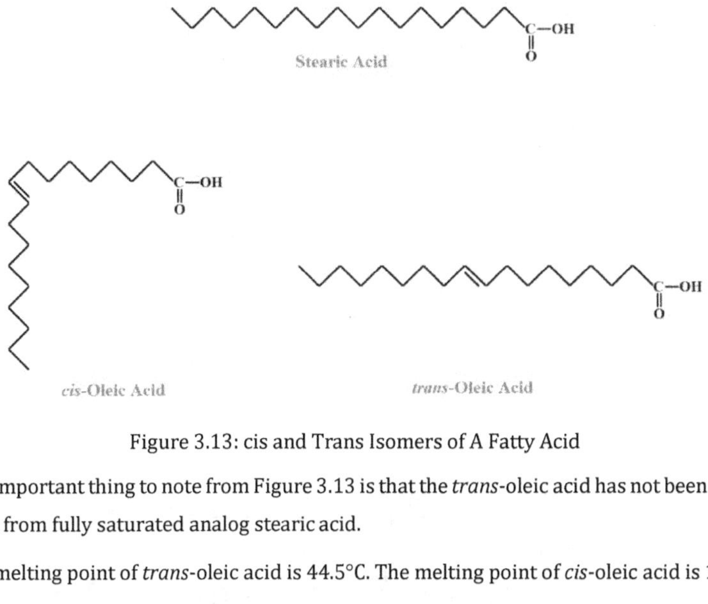

Figure 3.13: cis and Trans Isomers of A Fatty Acid

The important thing to note from Figure 3.13 is that the *trans*-oleic acid has not been changed in shape from fully saturated analog stearic acid.

The melting point of *trans*-oleic acid is 44.5°C. The melting point of *cis*-oleic acid is 13.4°C.

This difference is due to what is known as a **steric differences** between the two molecules. Molecules of *trans*-oleic acid can lay close to one another more easily than molecules of *cis*-oleic acid. Because of this, molecules of *trans*-oleic acid are more prone to be attracted to one another by intermolecular forces and will require more energy to melt and, thus a higher melting point.

Waxes

Most waxes are esters of saturated fatty acids and a long chained alcohol. Paraffin wax is comprised of long chained alkane hydrocarbons ranging from 20 to 40 carbons long.

Table 3.4 lists eight waxes in increasing order of melting points and lists the type and source of the waxes.

Table 3.4: Types and Source of Waxes

Wax	Type	Source	mp (°C)
Jojoba Oil	Plant	Jojoba Plant	7–11
Paraffin Wax	Hydrocarbon	Petroleum	37
Lanolin	Animal	Sheep's Wool	38
Spermaceti	Animal	Sperm Whale Oil	50
Beeswax	Animal	Bees	62–64
Candelilla	Plant	Candelilla Plant	69–73
Carnauba Wax	Plant	Brazilian Palm	82–85
Montan Wax	Hydrocarbon	Lignite Coal	82–95

3.4.   Lipids: Triacylglycerols to Glycerophospholipids

Triacylglycerols: Fats and Oils

Fatty acids are stored in the body as triacylglycerols, also known as triglycerides. These compounds are triesters of glycerol (a trihydroxy alcohol) and fatty acids.

Triacylglycerols have the general formulation shown in Figure 3.14.

$$H_2C-O-\overset{\overset{\displaystyle O}{\|}}{C}-(CH_2)_n-CH_3$$
$$HC-O-\overset{\overset{\displaystyle O}{\|}}{C}-(CH_2)_n-CH_3$$
$$H_2C-O-\overset{\overset{\displaystyle O}{\|}}{C}-(CH_2)_n-CH_3$$

Figure 3.14: A Triacylglycerol

In the drawing in Figure 3.14 the "n" represent the number of CH_2 groups in the long chain of the fatty acid. If the value of "n" were 16 in the triacylglycerides in Figure 3.14 the triacylglyceride would have been formed from glycerol and three molecules of stearic acid.

Fatty Acid Content of Fats and Oils

A **_fat_** is a triacylglycerol that is solid at room temperature and usually comes from animal sources. Some examples are lard, butter, and cheese.

An **_oil_** is a triacylglycerol that is liquid at room temperature and usually comes from plant sources. Some examples are peanut, safflower, and corn oils.

Table 3.5 lists saturated fatty acids (SFA), monounsaturated fatty acids (MUFA), polyunsaturated (PUFA) fatty acids content of some common fats and oils in descending order of percent MUFA content.

Table 3.5: Fatty Acid Content of Fats and Oils

Fat or Oil	% SFA	% MUFA	% PUFA
Olive	16	76	8
Safflower	12	75	13
Canola	6	62	32
Peanut	18	49	33
Beef Tallow	52	44	4
Soybean	14	40	46
Milk Butterfat	66	30	4
Walnut	13	25	62
Corn	17	24	59
Cottonseed	25	23	52
Palm Kernel	85	13	2
Coconut	91	6	3

The entries were listed in descending order of percent MUFA in order to list what this author considers the healthier oils first.

Effects on Melting Points of Fats and Oils

There are two major effects on the melting points of fats and oils. The first effect is the **shape** of the molecule. Notice the difference in the shapes and the melting points of the _cis_ and _trans_ isomers of oleic acid.

The _cis_ isomer is unable to lay in next to other _cis_ isomers therefore reducing the effects of intermolecular forces that act to hold _trans_ isomers together.

Figure 3.15 shows the _cis_ and _trans_ isomers of oleic acid and there melting points.

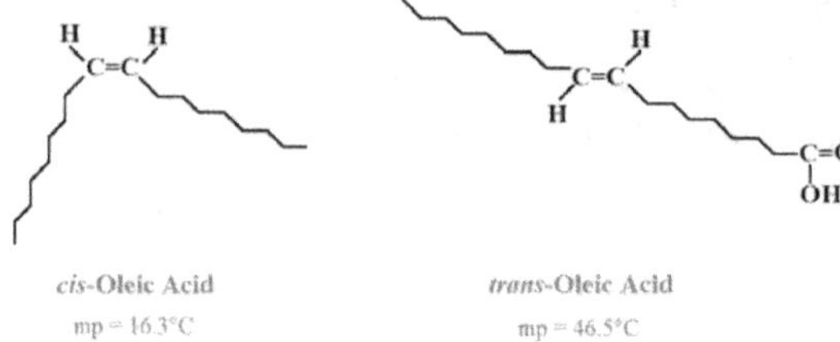

Figure 3.15: cis and trans Isomers of Oleic Acid

Note that the more linear *trans* isomer has the higher melting point.

The second effect is the ***degree of saturation*** in the fat or oil molecule. Notice the difference in the number of points of unsaturation and melting points of linoleic and stearic acid. Each molecule has the same number of carbon atoms.

Since linoleic acid has four less hydrogens than the stearic there is a pronounced effect on the amount of intermolecular forces and melting points of the two molecules.

Figure 3.16 shows the degrees of unsaturation and melting points for linoleic acid and stearic acid.

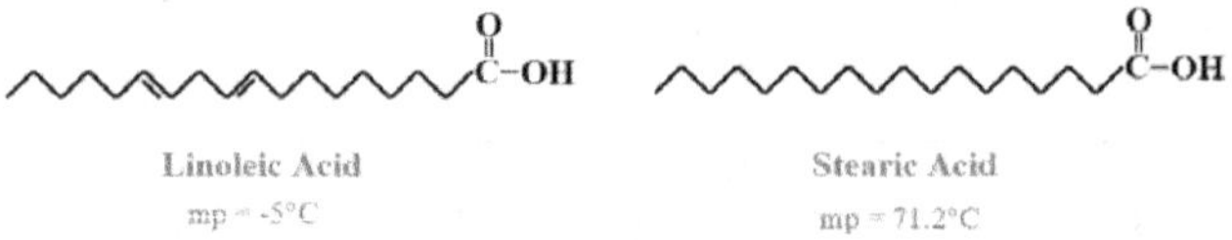

Figure 3.16: Linoleic Acid and Stearic Acid

Hydrogenation of Triacylglycerols

The **hydrogenation** of an unsaturated fat is accomplished by bubbling hydrogen gas through the fat in the presence of nickel.

Figure 3.17 shows glyceryltrioleate being converted to glyceryltristearate, a totally saturated fat.

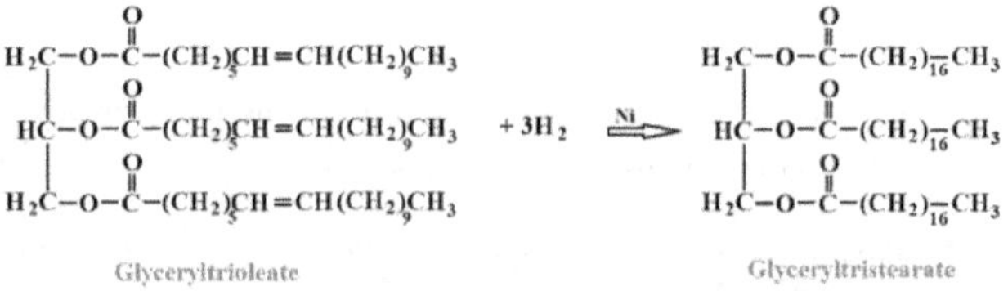

Figure 3.17: Hydrogenation of Glyceryltrioleate

Hydrolysis of Triacylglycerols

Triacylglycerols are **hydrolyzed** in the presence of strong acids or digestive enzymes called *lipases*. The resulting products are the original glycerol and the three original fatty acids.

Figure 3.18 shows the hydrolysis of glyceryltristearate to glycerol and three stearic acid molecules.

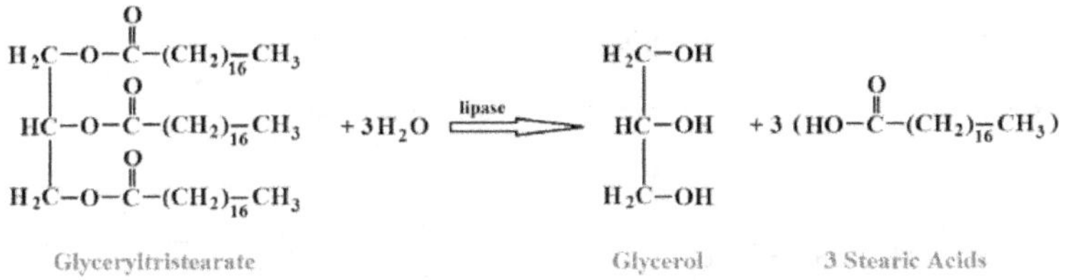

$$H_2C-O-\overset{O}{\overset{\|}{C}}-(CH_2)_{16}CH_3$$
$$HC-O-\overset{O}{\overset{\|}{C}}-(CH_2)_{16}CH_3 \quad +3H_2O \xrightarrow{\text{lipase}} \quad HC-OH \quad +3\,(HO-\overset{O}{\overset{\|}{C}}-(CH_2)_{16}CH_3)$$
$$H_2C-O-\overset{O}{\overset{\|}{C}}-(CH_2)_{16}CH_3 \qquad\qquad H_2C-OH$$

Glyceryltristearate   Glycerol   3 Stearic Acids

Figure 3.18: Hydrolysis of Glyceryltristearate

Saponification

When a fat is heated with a strong base such as sodium hydroxide (NaOH), **saponification** of the fat occurs yielding the original glycerol and three sodium esters of the original three fatty acids or **soap**.

Figure 3.19 shows the saponification of glyceryltristearate to glycerol and three sodium stearate molecules.

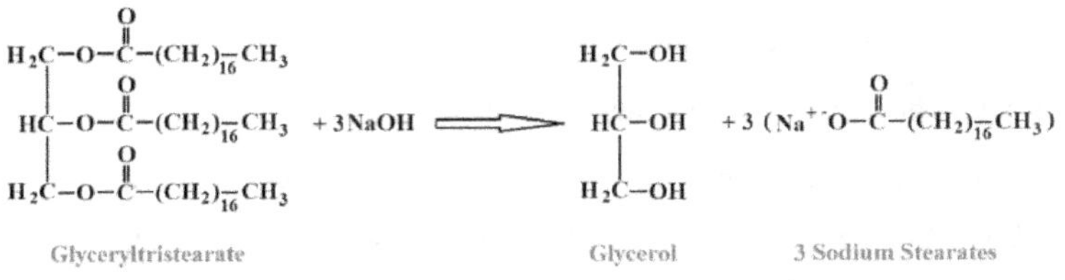

$$H_2C-O-\overset{O}{\overset{\|}{C}}-(CH_2)_{16}CH_3$$
$$HC-O-\overset{O}{\overset{\|}{C}}-(CH_2)_{16}CH_3 \quad +3NaOH \Longrightarrow \quad HC-OH \quad +3\,(Na^{+}\,{}^{-}O-\overset{O}{\overset{\|}{C}}-(CH_2)_{16}CH_3)$$
$$H_2C-O-\overset{O}{\overset{\|}{C}}-(CH_2)_{16}CH_3 \qquad\qquad H_2C-OH$$

Glyceryltristearate   Glycerol   3 Sodium Stearates

Figure 3.19: Saponification of Glyceryltristearate

Properties of Soaps

The soap product in the saponification reaction of Figure 3.19 would literally be your great grandparents' "lye soap".

If the sodium hydroxide (NaOH) in the reaction is replaced with potassium hydroxide (KOH) a softer soap would result. Polyunsaturated oils also produce softer soaps. If you see products such as "coconut or avocado" shampoos, you know the source of the oils used to make the soaps.

Glycerophospholipds

Glycerophospholipids are a family of lipids similar to triacylglycerols except that one of the fatty acid groups on the triacylglycerol has been replaced by the ester of phosphoric acid and an alcoholic amine.

Lecithins and *cephalins* are two types of glycerophospholipids that are abundant in brain and nerve tissue as well as the yolks of eggs. Lecithin contains the alcoholic amine *choline*. Cephalin contains the alcoholic amine *ethanolamine* and sometimes *serine*.

Figure 3.20 shows the glycerophospholipids lecithin and cephalin.

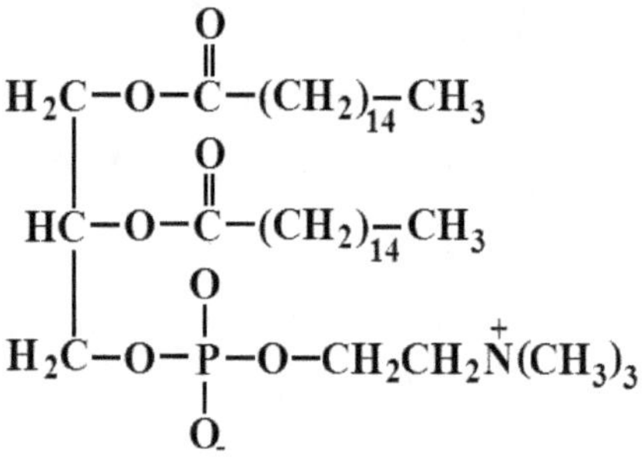

$$H_2C-O-\overset{\overset{O}{\|}}{C}-(CH_2)_{\overline{14}}CH_3$$
$$HC-O-\overset{\overset{O}{\|}}{C}-(CH_2)_{\overline{14}}CH_3$$
$$H_2C-O-\overset{\overset{O}{|}}{\underset{\underset{O^-}{|}}{P}}-O-CH_2CH_2\overset{+}{N}(CH_3)_3$$

Lecethin

$$H_2C-O-\overset{\overset{O}{\|}}{C}-(CH_2)_{\overline{14}}CH_3$$
$$HC-O-\overset{\overset{O}{\|}}{C}-(CH_2)_{\overline{14}}CH_3$$
$$H_2C-O-\overset{\overset{O}{|}}{\underset{\underset{O^-}{|}}{P}}-O-CH_2CH_2\overset{+}{N}H_3$$

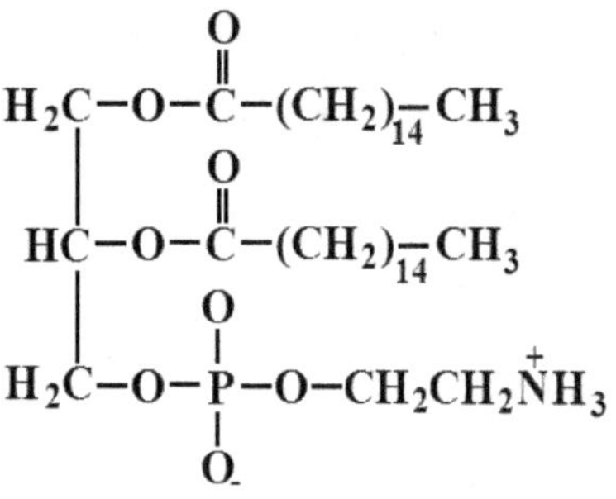

Cephalin

Figure 3.20: Lecithin and Cephalin

Biological Properties of Glycerophospholipids

Glycerophospholipids are the most abundant lipids in cell membranes.

Because they have both a polar (the ionized alcohol and phospho group)"head" and a nonpolar (fatty acid) "tail", they are able to "solubilize" the barely water soluble triglycerides and cholesterols and aid in their transport through the body.

Glycerophospholipids make up much of the myelin sheaths that protects nerve cells.

3.5. Steroids

Steroids are compounds comprising the steroid backbone which is composed of three fused cyclohexane rings fused to a cyclopentane ring.

Figure 3.21 shows the steroid backbone with the carbon numbering convention as well as individual ring designations.

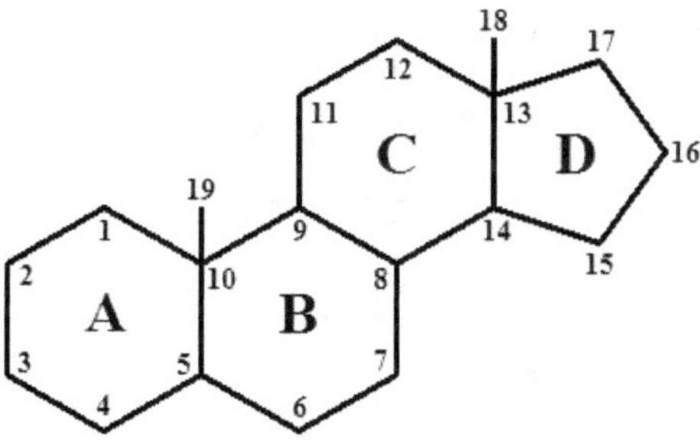

Figure 3.21: Steroid Backbone

Cholesterol

Cholesterol is a steroid referred to as a **sterol** due to the alcohol group on carbon 3. The methyl groups on carbons 10 and 13, the double bond between carbons 5 and 6, and the carbon chain on carbon 17 are typical of most steroids.

Figure 3.22 shows the cholesterol molecule.

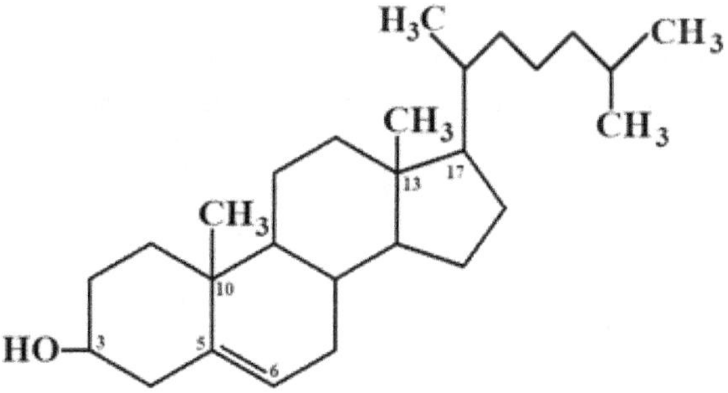

Figure 3.22: Cholesterol

The Role of Cholesterol in the Body

Cholesterol is a component of cell membranes, myelin sheath, and brain and nerve tissue. It is also found in the liver, bile salts, and skin.

If the diet is high in cholesterol then the liver produces less cholesterol.

The American Heart Association recommends that we consume no more than 300 mg cholesterol per day. Table 3.6 lists some common foodstuffs and their cholesterol content in mg cholesterol per 100 g of the foodstuff.

Table 3.6: Cholesterol Content of Common Foodstuffs

Foodstuff	mg/100 g Cholesterol
Eggs	1436
Butter	250
Cream Cheese	110
Animal Fat	95
Beef	72
Chicken	64
Whole Milk	11
Skim Milk	4
Fruits	0
Egg Whites	0
Grains	0
Nuts	0

Lipoproteins

Lipoproteins are spherical particles with a surface of polar proteins and glycerophospholipids surrounding hundreds of nonpolar triacylglycerols and cholesterol esters, the prevalent form of cholesterol in the blood.

Their function in the bloodstream is to **transport lipids** that are nonpolar and insoluble in the aqueous environment of the blood.

Figure 3.23 shows the structure of a lipoprotein.

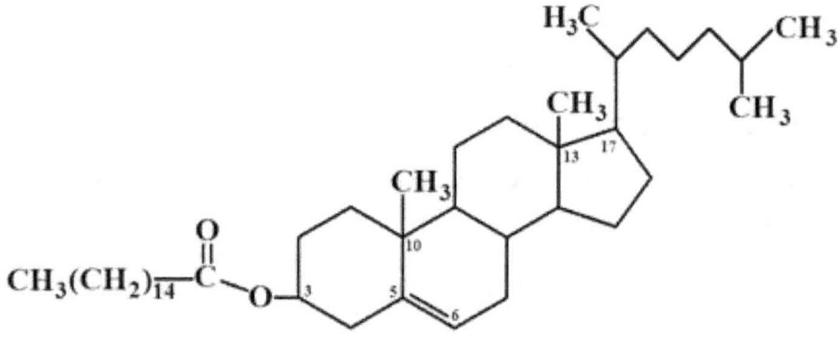

Figure 3.23: A Lipoprotein

Take note of the esterification of cholesterol (at carbon #3) that this lipoprotein represents.

Steroid Hormones

In the body hormones serve as a sort of communication system from one part of the body to another.

Steroid hormones include the sex hormones and adrenocortical hormones.

The sex hormones and adrenocortical hormones are closely related in structure to cholesterol and depend on cholesterol for their synthesis.

Figure 3.24 shows the structures of testosterone and estrogen.

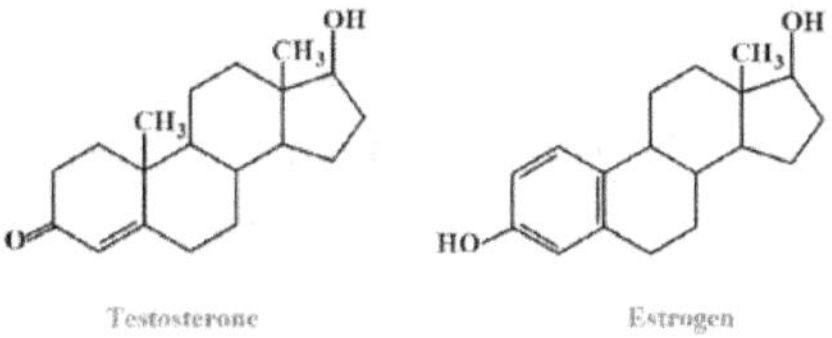

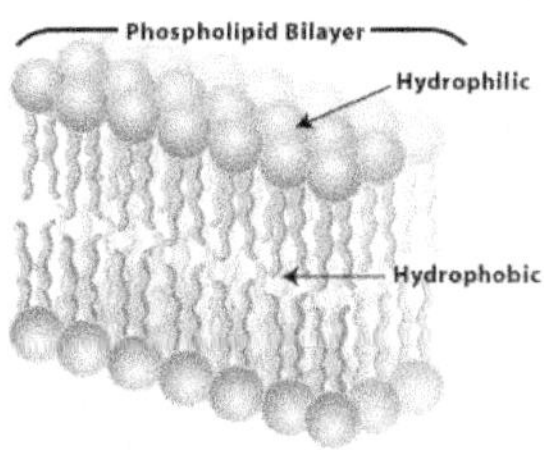

Figure 3.24: Testosterone and Estrogen

Notice how the term testosterone reflects the presence of a ketone group in the molecule.

The Lipid Bilayer of Cell Membranes

In a cell membrane two layers of glycerophospholipids are arranged like a sandwich.

Their nonpolar tails (which are **hydrophobic** or "water-fearing") are aligned to the center of the cell membrane while their polar heads (which are **hydrophilic** or "water-loving") are aligned to the outside of the cell membrane.

The hydrophobic layer aligns with the nonpolar contents of the cell while the hydrophilic layer aligns with the aqueous environment on the outer surface of the cell membrane in what is known as a **lipid bilayer**.

Figure 3.25 shows a cross section of a lipid bilayer.

Figure 3.25: A Lipid Bilayer

Questions

Part A

1. What are carbohydrates?
2. List the types of carbohydrates.
3. What are Polysaccharides?
4. Draw the Ring Structures of Monosaccharides
5. What is Cellulose?
6. Define Lipids
7. What are fatty acids?
8. What is meant by saponification ?
9. What are the Effects on Melting Points of Fats and Oils ?
10. Define Cholesterol

Part B

1. Give detailed explanation about Carbohydrates
2. Explain in detail about cellulose and glycogen
3. Briefly explain The Lipid Bilayer of Cell Membranes

CHAPTER 4

INORGANIC CHEMISTRY

Chemistry of Nonmetallic Elements

There are about 20 nonmetallic elements which are generally found as either anions in ionic compounds or else as elementary substances. It is possible to learn the names, structures, and main properties of these various compounds following a relatively simple classification. Hydrides, oxides, sulfides, and halides are important, and essential for the study of pure and applied inorganic chemistry of the solid state compounds.

4.1. Hydrogen and Hydrides

(a) Hydrogen

Hydrogen is the simplest element consisting of a proton and an electron, and the most abundant element in the universe. It is next to oxygen and silicon, and about 1 wt% of all the elements on the Earth. Needless to say, most hydrogen exists as water on the Earth. Since its polarity may change freely between hydride (H^-), atom (H), and proton (H^+), hydrogen also forms various compounds with many elements including oxygen and carbon. Therefore, hydrogen is highly important in chemistry.

Of the three kinds of isotopes of hydrogen, deuterium, D, was discovered by H.C. Urey and others in 1932, and subsequently tritium, T, was prepared from deuterium in 1934. About 0.015% of hydrogen is present as deuterium, and this can be enriched by electrolysis of water. Tritium is a radioactive isotope emitting β-particles with a half-life of 12.33 years. Since the mass of deuterium and tritium is about twice and three times that of hydrogen, respectively, the physical properties of the isotopes, and compounds containing them, differ considerably. Some properties of the isotopes and water are listed in Table 4.1. When the E-H bond in a hydrogen compound is converted into the E-D by deuterium substitution, the E-H stretching frequency in an infrared spectrum is reduced to about $1/\sqrt{2}$ which is useful for determining the position of the hydrogen atom. It is sometimes possible to conclude that scission of the bond with a hydrogen is the rate-determining step when the deuterium substitution shows a marked effect on the rate of reaction of a hydrogen-containing compound.

Since the nuclear spin of hydrogen is 1/2 and given its abundance, it is the most important nuclide for NMR spectroscopy. NMR is widely used not only for identification of organic compounds, but also for medical diagnostic purposes using MRI (magnetic resonance imaging) of water in living bodies. Human organs can now be observed with this non-invasive method.

Table 4.1: Properties of Isotopic Hydrogen and Water

Properties	H_2	D_2	T_2	H_2O	D_2O	T_2O
Melting point*	13.957	18.73	20.62	0.00	3.81	4.48
Boiling point	20.39	23.67	25.04	100.00	101.42	101.51
Density (g cm^{-3}, 25°C)				0.9970	1.1044	1.2138
Temp. of maximum density (°C)				3.98	11.23	13.4

* hydrogen (K), water (°C)

There are nuclear-spin isomers in diatomic molecules of the nuclides whose spin is not zero. Especially in the case of a hydrogen molecule, the difference of properties is significant. Spins of ***para*-hydrogen** are anti-parallel and the sum is 0 leading to a **singlet state**. Spins of ***ortho*-hydrogen** are parallel and the sum is 1 resulting in a **triplet state**. Since *para*-hydrogen is in a lower energy state, it is the stabler form at low temperatures. The theoretical ratio of *para*-hydrogen is 100% at 0 K, but it decreases to about 25% at room temperature, since the ratio of *ortho*-hydrogen increases at higher temperatures. Gas chromatography and rotational lines in the electronic band spectrum of H_2 can distinguish two hydrogen isomers.

(b) Hydride

Binary hydrides can be classified according to the position of the element in the periodic table, and by the bond characters. The hydrides of alkali and alkaline earth metals among *s*-block elements are ionic compounds structurally analogous to halides and are called **saline hydrides**. The Group 13-17 *p*-block elements form covalent molecular hydrides. No hydride of rare gas elements has been reported.

Some of the *d*-block and *f*-block transition metals form metal hydrides exhibiting metallic properties. Transition metals which do not give binary hydrides form many **molecular hydride complexes** coordinated by stabilization ligands, such as carbonyl (CO), tertiaryphosphines (PR$_3$), or cyclopentadienyl (C_5H_5). Typical hydrides of each class are given below.

Saline Hydrides

Lithium hydride, LiH, is a colorless crystalline compound (mp (melting point) 680°C). Li$^+$ and H$^-$ form a lattice with a rock salt type structure. Quantitative evolution of hydrogen gas at the anode during the electrolysis of the fused salt suggests the existence of H$^-$. Water reacts vigorously with lithium hydride evolving hydrogen gas. Since it dissolves in ethers slightly, the hydride is used as a reducing agent in organic chemistry.

Calcium hydride, CaH_2, is a colorless crystalline compound (mp 816°C), and reacts mildly with water evolving hydrogen gas. This hydride is used as a hydrogen gas generator, or a dehydrating agent for organic solvents. It is used also as a reducing agent.

Lithium tetrahydridoaluminate, $LiAlH_4$, is a colorless crystalline compound (decomposes above 125°C) usually called lithium aluminum hydride. The hydride dissolves in ethers, and reacts violently with water. It is used as a reducing and hydrogenating agent and for dehydrating organic solvents.

Sodium tetrahydroborate, $NaBH_4$, is a white crystalline compound (decomposes at 400°C) usually called sodium borohydride. It is soluble in water and decomposes at high temperatures evolving hydrogen gas. It is used as a reducing agent for inorganic and organic compounds, for preparation of hydride complexes, *etc.*

Molecular Hydrides

All hydrides other than those of carbon (methane) and oxygen (water) are poisonous gases with very high reactivity and should be handled very carefully. Although there are methods of generating the gases in laboratories, recently many are also available in cylinders.

Diborane, B_2H_6, is a colorless and poisonous gas (mp -164.9°C and bp -92.6°C) with a characteristic irritating odor. This hydride is a powerful reducing agent of inorganic and organic compounds. It is also useful in organic synthesis as a hydroboration agent that introduces functional groups to olefins, after addition of an olefin followed by reactions with suitable reagents.

Silane, SiH_4, is a colorless and deadly poisonous gas (mp -185°C and bp -111.9°C) with a pungent smell, and is called also monosilane.

Ammonia, NH_3, is a colorless and poisonous gas (mp -77.7°C and bp -33.4°C) with a characteristic irritating odor. Although it is used in many cases as aqueous ammonia since it dissolves well in water, liquid ammonia is also used as a nonaqueous solvent for special reactions. Since the Harber-Bosch process of ammonia synthesis was developed in 1913, it has been one of the most important compounds in chemical industries and is used as a starting chemical for many nitrogenous compounds. It is used also as a refrigerant.

Phosphine, PH_3, is a colorless and deadly poisonous gas (mp -133°C and bp -87.7°C) with a bad smell, and is called also phosphorus hydride. It burns spontaneously in air. It is used in vapor phase epitaxial growth, in transition metal coordination chemistry, *etc.*

Hydrogen sulfide, H_2S, is a colorless and deadly poisonous gas (mp -85.5°C and bp -60.7°C) with a rotten egg odor. Although often used with insufficient care, it is very dangerous and should be handled only in an environment with good ventilation. It is used in chemical analysis for the precipitation of metal ions, preparation of sulfur compounds, *etc.*

Hydrogen fluoride, HF, is a colorless, fuming, and low boiling point liquid (mp -83°C and bp 19.5°C), with an irritating odor. It is used for preparing inorganic and organic fluorine compounds. Because of its high permittivity, it can be used as a special nonaqueous solvent. The aqueous solution is called fluoric acid and is stored in polyethylene containers since the acid corrodes glass.

Metallic Hydrides

The hydrides MH_x which show metallic properties are nonstoichiometric interstitial-type solids in which hydrogen occupies a part of the cavities of the metal lattice. Usually x is not an integer in these compounds. There are Group 3 (Sc, Y), Group 4 (Ti, Zr, Hf), Group 5 (V, Nb, Ta), Cr, Ni, Pd, and Cu metallic hydrides among the *d* block elements, but the hydrides of other metals in Group 6 to 11 are not known. Palladium Pd reacts with hydrogen gas at ambient temperatures, and forms hydrides that have the composition PdH_x (x<1). Many metallic hydrides show metallic conductivity. $LaNi_5$ is an intermetallic compound of lanthanum and nickel. It occludes nearly 6 hydrogen atoms per unit lattice and is converted to $LaNi_5H_6$. It is one of the candidates for use as a hydrogen storage material with the development of hydrogen-fueled cars.

Problem 4.1: Write the oxidation number of the hydrogen atom in H_2, NaH, NH_3, and HCl.

[Answer] H_2 (0), NaH (-1), NH_3 (+1), and HCl (+1).

Hydride Complexes

Complexes coordinated by hydride ligands are called **hydride complexes**.

The Group 6 to 10 transition metals that do not form binary hydrides give many hydride complexes with auxiliary ligands such as carbonyl and tertiaryphosphines. Although it was only at the end of the 1950s that hydride was accepted as a ligand, thousands of hydride complexes are known at present. Furthermore, with the synthesis in the 1980's of molecular hydrogen complexes, the chemistry of transition metal hydrogen compounds took a new turn. Research on the homogeneous catalysis of hydrocarbons in which hydride or dihydrogen complexes participate is also progressing.

4.2. Main group Elements of 2nd and 3rd Periods and their Compounds

(a) Boron

Refined elemental boron is a black solid with a metallic luster. The unit cell of crystalline boron contains 12, 50, or 105 boron atoms, and the B_{12} icosahedral structural units join together by 2 center 2 electron (2c-2e) bonds and 3 center 2 electron (3c-2e) bonds (electron deficient bonds) between boron atoms (Figure 4.1). Boron is very hard and shows semiconductivity.

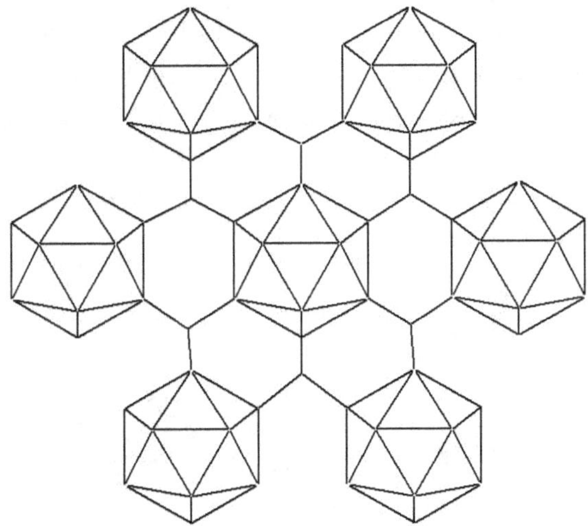

Figure 4.1: The Structure of the Elemental Boron with B_{12} Icosahedra

The chemistry of **boranes** (boron hydrides) started from the research of A. Stock reported during the period 1912-1936. Although boron is adjacent to carbon in the periodic table, its hydrides have completely different properties from those of hydrocarbons. The structures of boron hydrides in particular were unexpected and could be explained only by a new concept in chemical bonding. For his contribution to the very extensive new inorganic chemistry of boron hydrides, W. N. Lipscomb won the Nobel prize in 1976. Another Nobel prize (1979) was awarded to H. C. Brown for the discovery and development of a very useful reaction in organic synthesis called hydroboration.

Because of the many difficulties associated with the low boiling points of boranes, as well as their activity, toxicity, and air-sensitivity, Stock had to develop new experimental methods for handling the compounds *in vacuo*. Using these techniques, he prepared six boranes B_2H_6, B_4H_{10},

B_5H_9, B_5H_{11}, B_6H_{10}, and $B_{10}H_{14}$ by the reactions of magnesium boride, MgB_2, with inorganic acids, and determined their compositions. However, additional research was necessary to determine their structures. At present, the original synthetic method of Stock using MgB_2 as a starting compound is used only for the preparation of B_6H_{10}. Since reagents such as lithium tetrahydroborate, $LiBH_4$, and sodium tetrahydroborate, $NaBH_4$, are now readily available, and **diborane**, B_2H_6, is prepared according to the following equation, higher boranes are synthesized by the pyrolysis of diborane.

$$3LiBH_4 + 4BF_3.OEt_2 \rightleftharpoons 2B_2H_6 + 3LiBF_4 + 4Et_2O$$

A new theory of chemical bonding was introduced to account for the bonding structure of diborane, B_2H_6. Although an almost correct hydrogen-bridged structure for diborane was proposed in 1912, many chemists preferred an ethane-like structure, H_3B-BH_3, by analogy with hydrocarbons. However, H. C. Longuet-Higgins proposed the concept of the electron-deficient **3-center 2-electron bond** (3c-2e bond) and it was proven by electron diffraction in 1951 that the structure was correct (Figure 4.2).

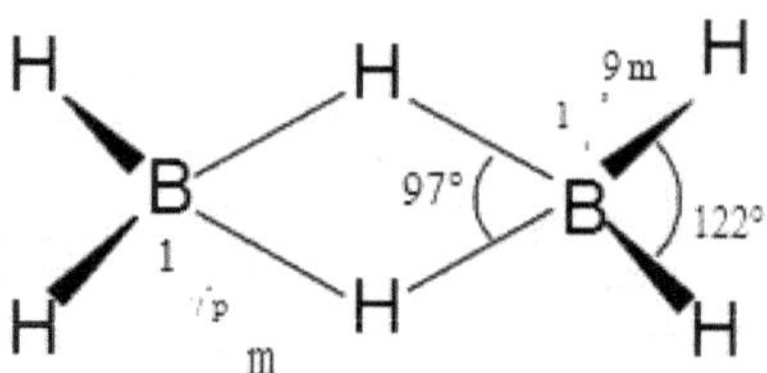

Figure 4.2: Structure of Diborane

It has been elucidated by electron diffraction, single crystal X-ray structure analysis, infrared spectroscopy, *etc.* that boranes contain 3-center 2-electron bonds (3c-2e bond) B-H-B and besides the usual 2 center 2 electron covalent bonds (2c-2e bond) B-H and B-B. Such structures can be treated satisfactorily by molecular orbital theory. Boranes are classified into *closo, nido, arachno, etc.* according to the skeletal structures of boron atoms.

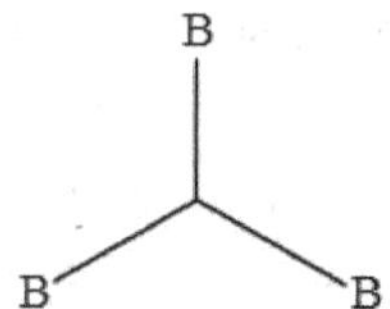

Closo-borane $[B_nH_n]^{2-}$ has the structure of a closed polyhedron of n boron atoms bonded to n hydrogen atoms, as seen in the examples of a regular octahedron $[B_6H_6]^{2-}$ and an icosahedron $[B_{12}H_{12}]^{2-}$. The boranes of this series do not contain B-H-B bonds. Boranes B_nH_{n+4}, such as B_5H_9, form structures with B-B, B-B-B, and B-H-B bonds and lack the apex of the polyhedron of *closo* boranes, and are referred to as **nido** type boranes. Borane B_nH_{n+6}, such as B_4H_{10}, have a structures that lacks two apexes from the *closo* type and are more open. Skeletons are also built by B-B, B-B-B, and B-H-B bonds, and these are called **arachno** type boranes. The structures of typical boranes are shown in Figure 4.3.

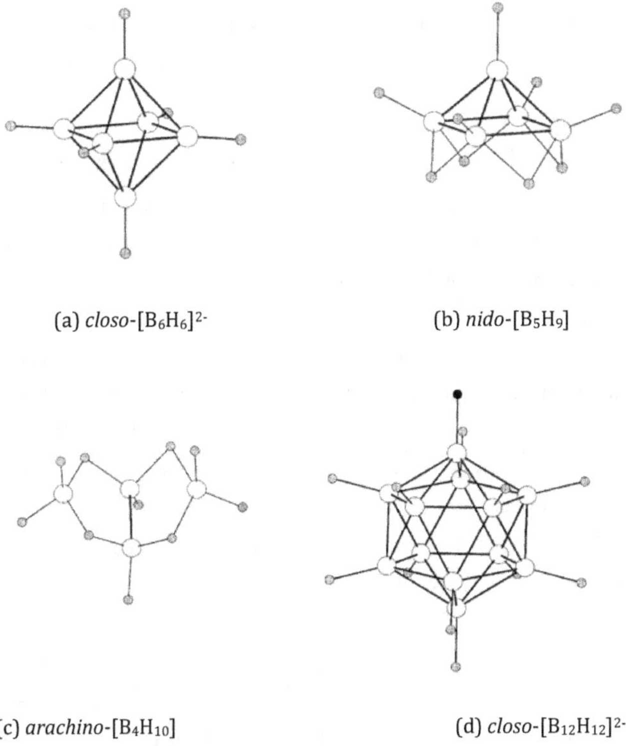

Figure 4.3: Structures of Boranes

Not only diborane but also higher boranes are **electron-deficient compounds** that are difficult to explain using Lewis' electronic structure based on simple 2-center 2-electron covalent bonds.

Problem 4.2: Why is diborane called an electron deficient compound?

[Answer] It is because there are only 12 valence electrons of boron and hydrogen atoms, although 16 electrons are necessary to assign two electrons each to eight B-H bonds.

K. Wade summarized the relation of the number of valence electrons used for skeletal bonds and the structures of boranes and proposed an empirical rule called the **Wade rule**. According to this rule, when the number of boron atoms is n, the skeletal valence electrons are $2(n+1)$ for a *closo* type, $2(n+2)$ for a *nido* type, and $2(n+3)$ for an *arachno* type borane. The relationship between the skeletal structure of a cluster compound and the number of valence electrons is also an important problem in the cluster compounds of transition metals, and the Wade rule has played a significant role in furthering our understanding of the structures of these compounds.

(b) Carbon

Graphite, diamond, fullerene, and amorphous carbon are carbon allotropes.

Usually a carbon atom forms four bonds using four valence electrons.

Graphite: Graphite is structured as layers of honeycomb-shaped 6 membered rings of carbon atoms that look like condensed benzene rings without any hydrogen atoms (Figure 4.4). The carbon-carbon distance between in-layer carbon atoms is 142 pm and the bonds have double bond character analogous to aromatic compounds. Since the distance between layers is 335 pm and the layers are held together by comparatively weak van der Waals forces, they slide when subjected to an applied force. This is the origin of the lubricating properties of graphite. Various molecules, such as alkali metals, halogens, metal halides, and organic compounds intercalate between the layers and form intercalation compounds. Graphite has semi-metallic electrical conductivity (about 10^{-3} Ωcm parallel to layers and about 100 times more resistant in the perpendicular direction).

Figure 4.4: Structure of Graphite

Diamond: Its structure is called the diamond-type structure (Figure 4.5). A unit cell of diamond contains eight carbon atoms and each carbon atom is 4-coordinate in a regular tetrahedron. Diamond is the hardest substance known, with a Mohs hardness 10. Diamond has very high heat conductivity although it is an electrical insulator. Although previously a precious mineral only formed naturally, industrial diamonds are now commercially prepared in large quantities at high temperatures (1200°C or higher) and under high pressures (5 GPa or more) from graphite using metal catalysts. In recent years, diamond thin films have been made at low temperatures (about 900°C) and under low pressures (about 102 Pa) by the pyrolysis of hydrocarbons, and are used for coating purposes, *etc.*

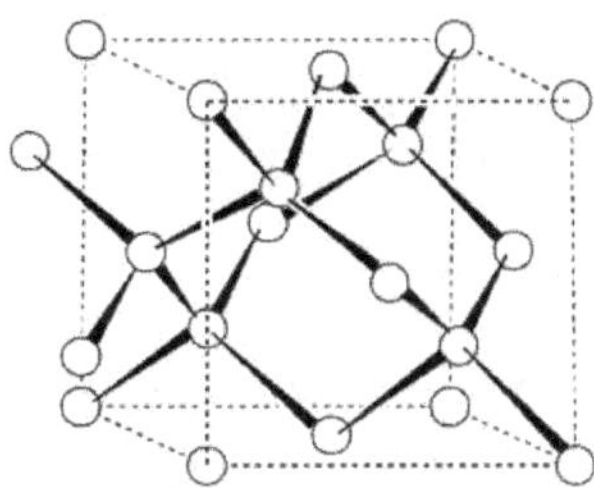

Figure 4.5: Structure of Diamond

Fullerene: Fullerene is the general name of the 3rd carbon allotrope, of which the soccer ball-shaped molecule C_{60} is a typical example (Figure 4.6). R. E. Smalley, H. W. Kroto and others detected C_{60} in the mass spectra of the laser heating product of graphite. In 1985, and fullerene's isolation from this so-called "soot" was reported in 1991. It has the structure of a truncated (corner-cut)-icosahedron and there is double bond character between carbon atoms. It is soluble in organic solvents, with benzene solutions being purple. Usually, it is isolated and purified by chromatography of fullerene mixtures. Wide-ranging research on chemical reactivities and physical properties such as superconductivity, is progressing rapidly. Besides C_{60}, C_{70} and carbon nanotubes are attracting interest.

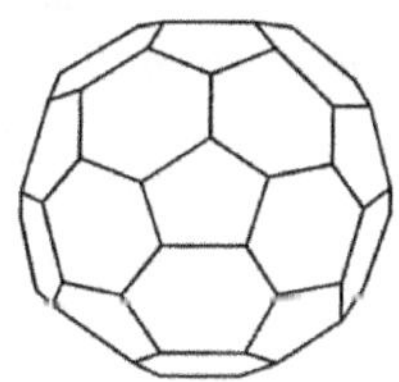

Figure 4.6: Structure of C_{60}

(c) Silicon

Silicon is the most abundant element in the earth's crust after oxygen. Most of this silicon exists as a component of silicate rocks and the element is not found as a simple substance. Therefore, silicon is produced by the reduction of quartz and sand with high-grade carbon using electric arc furnaces. Higher-grade silicon is obtained by hydrogen reduction of $SiHCl_3$, which is produced by the hydrochlorination of low purity silicon followed by rectification. The silicon used for semiconductor devices is further refined by the crystal Czochralski or zone melting methods. The crystal (mp 1410°C) has a metallic luster and the diamond type structure.

There are three isotopes of silicon, ^{28}Si (92.23%), ^{29}Si (4.67%), and ^{30}Si (3.10%). Because of its nuclear spin of $I = 1/2$, ^{29}Si is used for NMR studies of organic silicon compounds or silicates (solid-state NMR).

Silicates and organosilicon compounds show a wide range of structures in silicon chemistry. Section 4.3 (c) describes the properties of silicates. Organosilicon chemistry is the most active research area in the inorganic chemistry of main group elements other than carbon.

Silicon chemistry has progressed remarkably since the development of an industrial process to produce organosilicon compounds by the direct reaction of silicon with methyl chloride CH_3Cl in the presence of a copper catalyst. This historical process was discovered by E. G. Rochow in 1945. Silicone resin, silicone rubber, and silicone oil find wide application. In recent years, silicon compounds have also been widely used in selective organic syntheses.

Although silicon is a congener of carbon, their chemical properties differ considerably. A well-known example is the contrast of silicon dioxide SiO_2 with its 3-dimensional structure, and gaseous carbon dioxide, CO_2. The first compound $(Mes)_2Si=Si(Mes)_2$ (Mes is mesityl $C_6H_2(CH_3)_3$) with a silicon-silicon double bond was reported in 1981, in contrast with the ubiquitous carbon-carbon multiple bonds. Such compounds are used to stabilize unstable bonds with bulky substituents (kinetic stabilization).

Problem 4.3: Why are the properties of CO_2 and SiO_2 different?

[Answer] Their properties are very different because CO_2 is a chain-like three-atom molecule and SiO_2 is a solid compound with the three dimensional bridges between silicon and oxygen atoms.

(d) Nitrogen

Nitrogen is a colorless and odorless gas that occupies 78.1% of the atmosphere (volume ratio). It is produced in large quantities together with oxygen (bp -183.0°C) by liquefying air (bp -194.1°C) and fractionating nitrogen (bp -195.8°C). Nitrogen is an inert gas at room temperatures but converted into nitrogen compounds by biological nitrogen fixation and industrial ammonia synthesis. The cause of its inertness is the large bond energy of the N≡N triple bond.

The two isotopes of nitrogen are ^{14}N (99.634%) and ^{15}N (0.366%). Both isotopes are NMR-active nuclides.

(e) Phosphorus

Simple phosphorus is manufactured by the reduction of calcium phosphate, $Ca_3(PO_4)_2$, with quartz rock and coke. Allotropes include white phosphorus, red phosphorus, and black phosphorus.

White phosphorus is a molecule of composition of P_4 (Figure 4.7). It has a low melting-point (mp 44.1 °C) and is soluble in benzene or carbon disulfide. Because it is pyrophoric and deadly poisonous, it must be handling carefully.

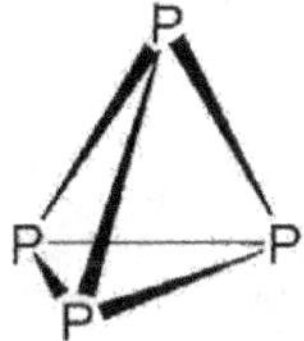

Figure 4.7: Structure of White Phosphorus

Red phosphorus is amorphous, and its structure is unclear. The principal component is assumed to be a chain formed by the polymerization of P_4 molecules as the result of the opening of one of the P-P bonds. It is neither pyrophoric nor poisonous, and used in large quantities for the manufacturing of matches, *etc.*

Black phosphorus is the most stable allotrope and is obtained from white phosphorus under high pressure (about 8 GPa). It is a solid with a metallic luster and a lamellar structure. Although it is a semiconductor under normal pressures, it shows metallic conductivity under high pressures (10 GPa).

Phosphorus Compounds as Ligands

Tertiary phosphines, PR_3, and phosphites, $P(OR)_3$, are very important ligands in transition metal complex chemistry. Especially triphenylphosphine, $P(C_6H_5)_3$, triethyl phosphine, $P(C_2H_5)_3$, and their derivatives are useful ligands in many complexes, because it is possible to control precisely their electronic and steric properties by modifying substituents. Although they are basically sigma donors, they can exhibit some pi accepting character by changing the substituents into electron accepting Ph (phenyl), OR, Cl, F, *etc.* The order of the electron-accepting character estimated from the C-O stretching vibrations and ^{13}C NMR chemical shifts of the phosphine- or phosphite-substituted metal carbonyl compounds is as follows (Ar is an aryl and R is an alkyl).

$$PF_3 > PCl_3 > P(OAr)_3 > P(OR)_3 > PAr_3 > PRAr_2 > PR_2Ar > PR_3$$

On the other hand, C. A. Tolman has proposed that the angle at the vertex of a cone that surrounds the substituents of a phosphorus ligand at the van der Waals contact distance can be a useful parameter to assess the steric bulkiness of phosphines and phosphites. This parameter, called the **cone angle,** is widely used (Figure 4.8). When the cone angle is large, the coordination number decreases by steric hindrance, and the dissociation equilibrium constant and dissociation rate of a phosphorus ligand become large (Table 4.2). The numerical expression of the steric effect is very useful, and many studies have been conducted into this effect.

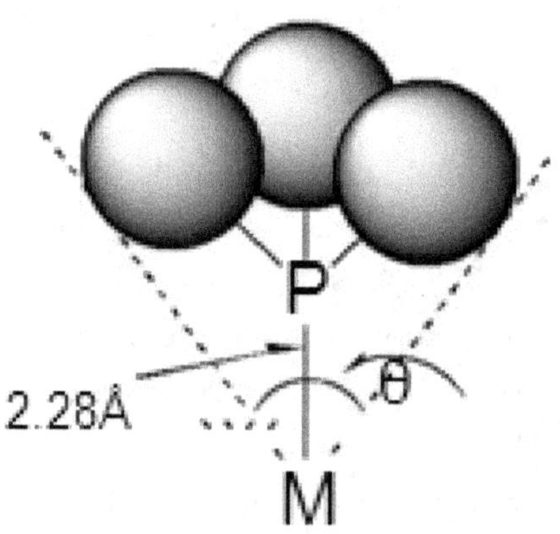

Figure 4.8: Cone Angle

Table 4.2: Cone angles ($\theta°$) of tertiary phosphines and phosphites

Ligands	Cone angles
P(OEt)$_3$	109
PMe$_3$	118
P(OPh)$_3$	121
PEt$_3$	132
PMe$_2$Ph	136
PPh$_3$	145
P^{i}Pr$_3$	160
P^{t}Bu$_3$	182

4.3. Oxygen and Oxides

(a) Oxygen

Dioxygen, O_2, is a colorless and odorless gas (bp-183.0°C) that occupies 21% of air (volume ratio). Since oxygen atoms are also the major components of water and rocks, oxygen is the most abundant element on the Earth's surface. Despite its abundance, it was established as an element as late as the 18th century. Since an immense quantity of oxygen gas is consumed for steel production now, it is separated in large quantities from liquified air.

The isotopes of oxygen are ^{16}O (99.762% abundance), ^{17}O (0.038%), and ^{18}O (0.200%). ^{17}O has nuclear spin $I = 5/2$ and is an important nuclide for NMR measurements. ^{18}O is used as a tracer for tracking reagents or for the study of reaction mechanisms. It is also useful for the assignment of absorption lines in infrared or Raman spectra by means of isotope effects.

As already described in section 2.3 (e), dioxygen, O_2, in the ground state has two unpaired spins in its molecular orbitals, shows paramagnetism and is called **triplet dioxygen**. In the excited state, the spins are paired and dioxygen becomes diamagnetic, which is called singlet dioxygen. Singlet dioxygen is important in synthetic chemistry, because it has characteristic oxidation reactivity. **Singlet dioxygen** is generated in a solution by an energy transfer reaction from a photo-activated complex or by the pyrolysis of ozonides (O_3 compounds).

Superoxide ion, O_2^-, and **peroxide ion**, O_2^{2-}, are the anions of dioxygen (Table 4.3). They can be isolated as alkali metal salts. There is another state, O_2^+, called the **dioxygen (1+) cation**, and it can be isolated as a salt with suitable anions.

Table 4.3: Oxidation States of Dioxygen

	Bond order	Compound	O-O distance (Å)	$(O\text{-}O)\ (cm^{-1})$
O_2^+	2.5	$O_2[AsF_6]$	1.123	1858
O_2	2.0		1.207	1554
O_2^-	1.5	$K[O_2]$	1.28	1145
O_2^{2-}	1.0	$Na_2[O_2]$	1.49	842

Ozone, O_3, is an allotrope of oxygen that is an unstable gas with an irritating odor. Ozone is a bent three-atom molecule (117°) and has unique reactivities. In recent years it has been discovered that ozone plays an important role in intercepting the detrimental ultraviolet radiation from the sun in the upper atmospheric zone, and in protecting life on the Earth from photochemical damage. It is now clear that chlorofluorocarbons, frequently used as refrigerants or as cleaners of electronic components, destroy the ozone layer, and measures are being taken on a global scale to cope with this serious environmental problem.

(b) Oxides of Hydrogen

Oxygen is highly reactive, and direct reactions with many elements form oxides.

Water is an oxide of hydrogen and is crucially important for the global environment and life in general.

Water H_2O

Ninety-seven percent of water on the Earth is present as sea water, 2% as ice of the polar zone, and fresh water represents only the small remaining fraction. Fundamental chemical and physical properties of water are very significant to chemistry. The main physical properties are shown in Table 4.1. Most of the unusual properties of water are caused by its strong hydrogen bonds. Physical properties of water differ considerably with the presence of isotopes of hydrogen. At least nine polymorphs of ice are known and their crystal structures depend on the freezing conditions of the ice.

Water has a bond angle of 104.5° and a bond distance of 95.7 pm as a free molecule. It is described in Section 3.4 (b) that self-dissociation of water generates oxonium ion, H_3O^+. Further water molecules add to H_3O^+ to form $[H(OH_2)_n]^+$ ($H_5O_2^+$, $H_7O_3^+$, $H_9O_4^+$, and $H_{13}O_6^+$), and the structures of the various species have been determined.

Hydrogen Peroxide H_2O_2

Hydrogen peroxide is an almost colorless liquid (mp -0.89 °C and bp (extrapolated) 151.4 °C) that is highly explosive and dangerous in high concentrations. Usually it is used as a dilute solution but occasionally 90% aqueous solutions are used. Since it is consumed in large quantities as a bleaching agent for fiber and paper, large-scale industrial synthetic process has been established. This process applies very subtle catalytic reactions to produce a dilute solution of hydrogen peroxide from air and hydrogen using a substituted anthraquinone. This dilute solution is then concentrated.

When deuterium peroxide is prepared in a laboratory, the following reaction is applied.

$$K_2S_2O_8 + 2\,D_2O \longrightarrow D_2O + 2\,KDSO_4$$

Hydrogen peroxide is decomposed into oxygen and water in the presence of catalysts such as manganese dioxide, MnO_2. Hydrogen peroxide may be either an oxidant or a reductant depending on its co-reactants. Its reduction potential in an acidic solution expressed in a Latimer diagram (refer to Section 3.3 (c)) is

$$O_2 \xrightarrow{\;+0.70\;} H_2O_2 \xrightarrow{\;+1.76\;} H_2O$$

(c) Silicon Oxides

Silicon oxides are formed by taking SiO_4 tetrahedra as structural units and sharing the corner oxygen atoms. They are classified by the number of corner-sharing oxygen atoms in the SiO_4 tetrahedra, as this determines their composition and structure. When the SiO_4 tetrahedra connect by corner sharing, the structures of the polymeric compounds become a chain, a ring, a layer, or 3-dimensional depending on the connection modes of adjacent units. Fractional expression is adopted in order to show the bridging modes. Namely, the numerator in the fraction is the number of bridging oxygens and the denominator is 2, meaning that one oxygen atom is shared by two tetrahedra. The empirical formulae are as follows and each structure is illustrated in Figure 4.9 in coordination-polyhedron form.

A bridge is constructed with one oxygen atom. $(SiO_3O_{1/2})^{3-} = Si_2O_7^{6-}$

Bridges are constructed with two oxygen atoms. $(SiO_2O_{2/2})_n^{2n-} = (SiO_3)_n^{2n-}$

Bridges are constructed with three oxygen atoms. $(SiOO_{3/2})_n^{n-} = (Si_2O_5)_n^{2n-}$

Amalgamation of bridging modes with three oxygen and two oxygen atoms.

$$[(Si_2O_5)(SiO_2O_{2/2})_2]_n^{6-} = (Si_4O_{11})_n^{6-}$$

Bridges are constructed with four oxygen atoms. $(SiO_{4/2})_n = (SiO_2)_n$

Silicates with various cross linkage structures are contained in natural rocks, sand, clay, soil, *etc.*

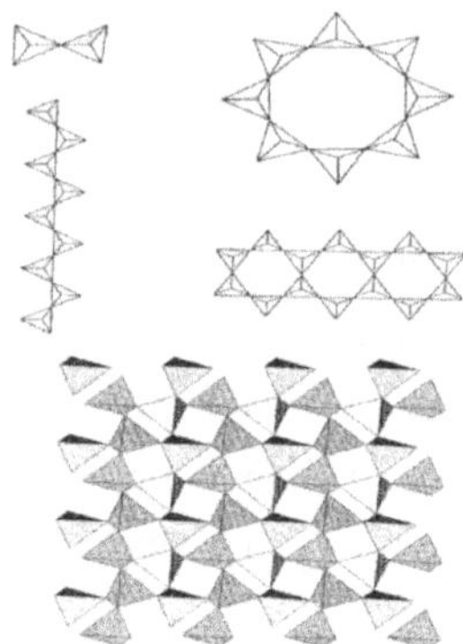

Figure 4.9: Bridging Modes of the SiO$_4$ Tetrahedra

Aluminosilicates

There are many minerals in which some silicon atoms of silicate minerals are replaced by aluminum atoms. They are called **aluminosilicates**. Aluminum atoms replace the silicon atoms in the tetrahedral sites or occupy the octahedral cavities of oxygen atoms, making the structures more complicated. The substitution of a tetravalent silicon by a trivalent aluminum causes a shortage of charge which is compensated by occlusion of extra cations such as H^+, Na^+, Ca^{2+}, *etc*. Feldspars are a typical aluminosilicate mineral, and $KAlSi_3O_8$ (orthoclase) and $NaAlSi_3O_8$ (albite) are also known well. Feldspars take 3-dimensional structures in which all the corners of the SiO_4 and AlO_4 tetrahedra are shared.

On the other hand, 2-dimensional layers are formed if $[AlSiO_5]^{3-}$ units are lined, and stratified minerals like mica are constructed if 6-coordinate ions are inserted between layers. If the number of oxygen atoms in the layers is not enough to form regular octahedra between layers, hydroxide groups bond to the interstitial Al^{3+} ions. Muscovite, $KAl_2(OH)_2Si_3AlO_{10}$, is a type of mica with such a structure and can be easily peeled into layers.

Zeolite

One of the important aluminosilicates is zeolite. Zeolites are present as natural minerals and also many kinds of zeolites are prepared synthetically in large quantities. The SiO_4 and AlO_4 tetrahedra are bonded by oxygen bridges, and form holes and tunnels of various sizes. The structures are composites of the basic structural units of tetrahedral MO_4. As shown in Figure 4.10, the basic units are cubes with 8 condensed MO_4, hexagonal prisms with 12 condensed MO_4, and truncated octahedra with 24 condensed MO_4.

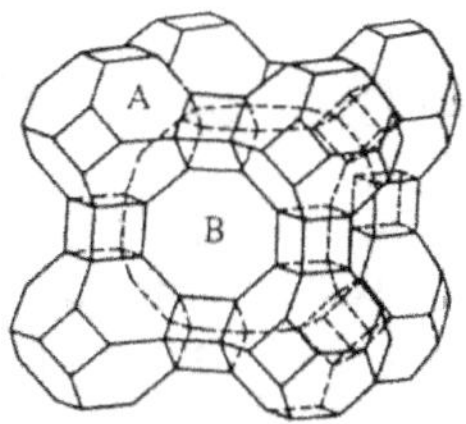

Figure 4.10: Structure of Zeolite A

Silicon or aluminum atoms are located on the corners of the polyhedra and the bridging oxygen atoms on the middle of each edge (it should be noted that this expression is different from the polyhedron model of oxides).

When these polyhdra are bonded, various kinds of zeolite structures are formed. For example, the truncated octahedra called β cages are the basic frames of **synthetic zeolite A**, $Na_{12}(Al_{12}Si_{12}O_{48})].27H_2O$, and the quadrangle portions are connected through cubes. It can be seen that an octagonal tunnel B forms when eight truncated octahedra bind in this way. The structure in which the hexagon portions connect through hexagonal prisms is faujasite, $NaCa_{0.5}(Al_2Si_5O_{14})] .10H_2O$.

Alkali metal or alkaline earth metal cations exist in the holes, and the number of these cations increases with the content of aluminum to compensate for the charge deficiency. The structures of zeolites have many crevices in which cations and water are contained. Utilizing this cation-exchange property, zeolites are used in large quantities as softeners of hard water. As zeolites dehydrated by heating absorb water efficiently, they are also used as desiccants of solvents or gases. Zeolites are sometimes called **molecular sieves**, since the sizes of holes and tunnels change with the kinds of zeolites and it is possible to segregate organic molecules according to their sizes. Zeolites can fix the directions of more than two molecules in their cavities and can be used as catalysts for selective reactions.

For example, synthetic zeolite ZSM-5 is useful as a catalyst to convert methanol to gasoline. This zeolite is prepared hydrothermally in an autoclave (high-pressure reaction vessel) at *ca.* 100 ºC using meta-sodium aluminate, $NaAlO_2$, as the source of aluminum oxide and silica sol as the source of silicon oxide and with tetrapropylammonium bromide, Pr_4NBr, present in the reaction. The role of this ammonium salt is a kind of mold to form zeolite holes of a fixed size. When the ammonium salt is removed by calcination at 500°C, the zeolite structure remains.

(d) Nitrogen Oxides

A variety of nitrogen oxides will be described sequentially from lower to higher oxidation numbers (Table 4.4).

Table 4.4: Typical Oxides of Main Group Elements

	1	2	12	13	14	15	16	17	18
2	Li_2O	BeO		B_2O_3	CO CO_2	N_2O NO NO_2			
3	Na_2O Na_2O_2 NaO_2	MgO		Al_2O_3	SiO_2	P_4O_6 P_4O_{10}	SO_2 SO_3	Cl_2O ClO_2	
4	K_2O K_2O_2 KO_2	CaO	ZnO	Ga_2O_3	GeO_2	As_4O_6 As_4O_{10}	SeO_2 SeO_3		
5	Rb_2O Rb_2O_2 Rb_9O_2	SrO	CdO	In_2O_3	SnO_2	Sb_4O_6 Sb_4O_{10}	TeO_2 TeO_3	I_2O_5	XeO_3 XeO_4
6	Cs_2O $Cs_{11}O_3$	BaO	HgO	Tl_2O Tl_2O_3	PbO PbO_2	Bi_2O_3			

Dinitrogen monoxide, N_2O. Oxide of monovalent nitrogen. Pyrolysis of ammonium nitrate generates this oxide as follows.

$$NH_4NO_3 \xrightarrow{250°C} N_2O + 2\,H_2O$$

Although the oxidation number is a formality, it is an interesting and symbolic aspect of the versatility of the oxidation number of nitrogen that NH_4NO_3 forms a monovalent nitrogen oxide (+1 is a half of the average of -3 and +5 for NH_4 and NO_3, respectively). The N-N-O bond distances of the straight N_2O are 112 pm (N-N) and 118 pm (N-O), corresponding to 2.5th and 1.5th bond order, respectively. N_2O (16e) is isoelectronic with carbon dioxide CO_2(16e). This compound is also called laughing gas and is widely used for analgesia.

Nitric oxide, NO. An oxide of divalent nitrogen. This is obtained by reduction of nitrite as follows.

$$KNO_2 + KI + H_2SO_4 \rightarrow NO + K_2SO_4 + H_2O + \frac{1}{2}\,I_2$$

Having an odd number of valence electrons (11 electrons), it is paramagnetic. The N-O distance is 115 pm and the bond has double bond character. The unpaired electron in the highest antibonding π^* orbital is easily removed, and NO becomes NO^+ (nitrosonium), which is isoelectronic with CO. Since an electron is lost from the antibonding orbital, the N-O bond becomes stronger. The compounds $NOBF_4$ and $NOHSO_4$ containing this cation are used as 1 electron oxidants.

Although NO is paramagnetic as a monomer in the gas phase, dimerization in the condensed phase leads to diamagnetism. It is unique as a ligand of transition metal complexes and forms complexes like $[Fe(CO)_2(NO)_2]$, in which NO is a neutral 3-electron ligand. Although M-N-O is straight in these kind of complexes, the M-N-O angle bends to $120°\sim140°$ in $[Co(NH_3)_5NO]$ Br_2, in which NO^- coordinates as a 4-electron ligand. It has become clear recently that nitric oxide has various biological control functions, such as blood-pressure depressing action, and it attracts attention as the second inorganic material after Ca^{2+} to play a role in signal transduction.

Dinitrogen trioxide, N_2O_3. The oxidation number of nitrogen is +3, and this is an unstable compound decomposing into NO and NO_2 at room temperature. It is generated when equivalent quantities of NO and NO_2 are condensed at low temperatures. It is light blue in the solid state and dark blue in the liquid state but the color fades at higher temperatures.

Nitrogen dioxide, NO_2. A nitrogen compound with oxidation number +4. It is an odd electron compound with an unpaired electron, and is dark reddish brown in color. It is in equilibrium with the colorless dimer dinitrogen tetroxide, N_2O_4. The proportion of NO_2 is 0.01% at -11°C, and it increases gradually to 15.9% at its boiling point (21.2°C), and becomes 100% at 140°C. N_2O_4 can be generated by the pyrolysis of lead nitrate as follows.

$$2\,Pb(NO_3)_2 \xrightarrow{\hspace{1cm}} 4\,NO_2 + 2\,PbO + O_2$$
$$400°C$$

When NO_2 is dissolved in water, nitric acid and nitrous acid are formed.

$$2\,NO_2 + H_2O \xrightarrow{\hspace{1cm}} HNO_3 + HNO_2$$

By one electron oxidation, NO^{2+} (nitroyl) forms and the O-N-O angle changes from 134° to 180° in the neutral NO_2. On the other hand, by one electron reduction, NO^{2-} (nitrito) forms and the angle bends to 115°.

Dinitrogen pentoxide, N_2O_5, is obtained when concentrated nitric acid is carefully dehydrated with phosphorus pentoxide at low temperatures. It sublimes at 32.4 °C. As it forms nitric acid by dissolving in water, it may also be called a nitric anhydride.

$$N_2O_5 + H_2O \longrightarrow 2\,HNO_3$$

Although it assumes an ion-pair structure NO_2NO_3 and straight NO_2^+ and planar NO_3^- ions are located alternately in the solid phase, it is molecular in the gas phase.

Oxoacids

Oxyacids of nitrogen include nitric acid, HNO_3, nitrous acid, HNO_2, and hyponitrous acid, $H_2N_2O_2$. Nitric acid, HNO_3, is one of the most important acids in the chemical industry, along with sulfuric acid and hydrochloric acid. Nitric acid is produced industrially by the **Ostwald process**, which is the oxidation reaction of ammonia in which the oxidation number of nitrogen increases from -3 to +5. Because the Gibbs energy of the direct conversion of dinitrogen to the intermediate NO_2 is positive, and therefore the reaction is unfavorable thermodynamically, dinitrogen is firstly reduced to ammonia, and this is then oxidized to NO_2.

$$N_2 \longrightarrow NH_3 \longrightarrow NO_2 \longrightarrow HNO_3$$

$$0 \qquad\qquad -3 \qquad\qquad +4 \qquad\qquad +5$$

Nitric acid, HNO_3. Commercial nitric acid is a *ca.*70% aqueous solution and vacuum distillation of it in the presence of phosphorus pentoxide gives pure nitric acid. As it is a strong oxidizing agent while also being a strong acid, it can dissolve metals (copper, silver, lead, *etc.*) which do not dissolve in other acids. Gold and platinum can even be dissolved in a mixture of nitric acid and hydrochloric acid (*aqua regia*). The nitrate ion, NO_3^-, and nitrite ion, NO_2^-, take various coordination forms when they coordinate as ligands in transition metal complexes.

Nitrous acid, HNO_2. Although not isolated as a pure compound, aqueous solutions are weak acids (pKa = 3.15 at 25 ℃) and important reagents. Since $NaNO_2$ is used industrially for hydroxylamine (NH_2OH) production and also used for diazotidation of aromatic amines, it is important for the manufacture of azo dyes and drugs. Among the various coordination forms of NO_2^- isomers now known, monodentate nitro (N-coordination) and nitrito (O-coordination) ligands had already been discovered in the 19th century.

(e) Phosphorus Oxides

The structures of the phosphorus oxides P_4O_{10}, P_4O_9, P_4O_7, and P_4O_6 have been determined.

Phosphorus pentoxide, P_4O_{10}, is a white crystalline and sublimable solid that is formed when phosphorus is oxidized completely. Four phosphorus atoms form a tetrahedron and they are bridged by oxygen atoms. Since a terminal oxygen atom is bonded to each phosphorus atom, the coordination polyhedron of oxygen is also a tetrahedron. When the molecular P_4O_{10}

is heated, a vitrified isomer is formed. This is a polymer composed of similar tetrahedra of phosphorus oxide with the same composition that are connected to one another in sheets. Since it is very reactive with water , phosphorus pentoxide is a powerful dehydrating agent. It is used not only as a desiccant, but also it has remarkable dehydration properties, and N_2O_5 or SO_3 can be formed by dehydration of HNO_3 or H_2SO_4, respectively. Phosphorus pentoxide forms orthophosphoric acid, H_3PO_4, when reacted with sufficient water, but if insufficient water is used, various kinds of condensed phosphoric acids are produced depending on the quantity of reacting water.

Phosphorus trioxide, P_4O_6, is a molecular oxide, and its tetrahedral structure results from the removal of only the terminal oxygen atoms from phosphorus pentoxide.

Each phosphorus is tri-coordinate. This compound is formed when white phosphorus is oxidized at low temperatures in insufficient oxygen. The oxides with compositions intermediate between phosphorus pentoxide and trioxide have 3 to 1 terminal oxygen atoms and their structures have been analyzed.

Although arsenic and antimony give molecular oxides As_4O_6 and Sb_4O_6 that have similar structures to P_4O_6, bismuth forms a polymeric oxide of composition Bi_2O_3.

Phosphoric Acid

Orthophosphoric acid, H_3PO_4. It is one of the major acids used in chemical industry, and is produced by the hydration reaction of phosphorus pentoxide, P_4O_{10}. Commercial phosphoric acid is usually of 75-85% purity. The pure acid is a crystalline compound (mp 42.35 $^\circ$C). One terminal oxygen atom and three OH groups are bonded to the phosphorus atom in the center of a tetrahedron. The three OH groups release protons making the acid tribasic ($pK_1 = 2.15$). When two orthophosphoric acid molecules condense by the removal of an H_2O molecule, pyrophosphoric acid, $H_4P_2O_7$, is formed.

Phosphonic acid, H_3PO_3. This acid is also called phosphorous acid and has H in place of one of the OH groups of orthophosphoric acid. Since there are only two OH groups, it is a dibasic acid.

Phosphinic acid, H_3PO_2. It is also called hypophosphorous acid, and two of the OH groups in orthophosphoric acid are replaced by H atoms. The remaining one OH group shows monobasic acidity. If the PO_4 tetrahedra in the above phosphorus acids bind by O bridges, many condensed phosphoric acids form. Adenosine triphosphate (ATP), deoxyribonucleic acid

(DNA), *etc.*, in which the triphosphorus acid moieties are combined with adenosine are phosphorus compounds that are fundamentally important for living organisms.

(f) Sulfur Oxides

Sulfur dioxide, SO_2. This is formed by the combustion of sulfur or sulfur compounds. This is a colorless and poisonous gas (bp -10.0 ^oC) and as an industrial emission is one of the greatest causes of environmental problems. However, it is very important industrially as a source material of sulfur. Sulfur dioxide is an angular molecule, and recently it has been demonstrated that it takes various coordination modes as a ligand to transition metals. It is a nonaqueous solvent similar to liquid ammonia, and is used for special reactions or as a solvent for special NMR measurements.

Sulfur trioxide, SO_3. It is produced by catalytic oxidation of sulfur dioxide and used for manufacturing sulfuric acid. The usual commercial reagent is a liquid (bp 44.6^oC). The gaseous phase monomer is a planar molecule. It is in equilibrium with a ring trimer (γ-SO_3 = S_3O_9) in the gaseous or liquid phase. In the presence of a minute amount of water SO_3 changes to β-SO_3, which is a crystalline high polymer with a helical structure. α-SO_3 is also known as a solid of still more complicated lamellar structure. All react violently with water to form sulfuric acid.

Sulfur Acids

Although there are many oxy acids of sulfur, most of them are unstable and cannot be isolated. They are composed of a combination of S=O, S-OH, S-O-S, and S-S bonds with a central sulfur atom. As the oxidation number of sulfur atoms varies widely, various redox equilibria are involved.

Sulfuric acid, H_2SO_4. It is an important basic compound produced in the largest quantity of all inorganic compounds. Pure sulfuric acid is a viscous liquid (mp 10.37^oC), and dissolves in water with the generation of a large amount of heat to give strongly acidic solutions.

Thiosulfuric acid, $H_2S_2O_3$. Although it is generated if thiosulfate is acidified, the free acid is unstable. The $S_2O_3^{2-}$ ion is derived from the replacement of one of the oxygen atoms of SO_4^{2-} by sulfur, and is mildly reducing.

Sulfurous acid, H_2SO_3. The salt is very stable although the free acid has not been isolated. The SO_3^{2-} ion that has pyramidal C_{3v} symmetry is a reducing agent. In dithionic acid, $H_2S_2O_6$, and the dithionite ion, $S_2O_6^{2-}$, the oxidation number of sulfur is +5, and one S-S bond is formed. This is a very strong reducing agent.

(g) Metal Oxides

Oxides of all the metallic elements are known and they show a wide range of properties in terms of structures, acidity and basicity, and conductivity. Namely, an oxide can exhibit molecular, 1-dimensional chain, 2-dimensional layer, or 3-dimensional structures. There are basic, amphoteric, and acidic oxides depending on the identity of the metallic element. Moreover, the range of physical properties displayed is also broad, from insulators, to semiconductors, metallic conductors, and superconductors. The compositions of metallic oxide can be simply stoichiometric, stoichiometric but not simple, or sometimes non-stoichiometric. Therefore, it is better to classify oxides according to each property. However, since structures give the most useful information to understand physical and chemical properties, typical oxides are classified first according to the dimensionality of their structures (Table 4.4, Table 4.5).

Table 4.5: Typical Binary Oxides of Transition Metals

Oxidation number	3	4	5	6	7	8	9	10	11
+1		Ti_2O^l							Cu_2O Ag_2O
+2		TiO	VO NbO		MnO	FeO	CoO	NiO	CuO Ag_2O_2
+3	Sc_2O_3 Y_2O_3	Ti_2O_3	V_2O_3	Cr_2O_3		Fe_2O_3	Rh_2O_3		
+4		TiO_2 ZrO_2 HfO_2	VO_2 NbO_2 TaO_2	CrO_2 MoO_2 WO_2	MnO_2 TcO_2 ReO_2	RuO_2 OsO_2	RhO_2 IrO_2	PtO_2	
+5			$V_2O_5^l$ Nb_2O_5 Ta_2O_5						
+6				CrO_3^c MoO_3^l WO_3	ReO_3				
+7					$Re_2O_7^l$				
+8						RuO_4^m OsO_4^m			

^{m} molecular, ^{c} chain, ^{l} layer, others 3-demensional.

Molecular Oxides

Ruthenium tetroxide, RuO_4, (mp 25°C and bp 40°C) and osmium tetroxide, OsO_4, (mp 40°C and bp 130°C) have low melting and boiling points and their structures are molecular. They are prepared by heating the metal powder in an oxygen atmosphere at about 800°C. The structures are tetrahedral and they are soluble in organic solvents and also slightly soluble in water. OsO_4 is used in organic chemistry especially in the preparation of *cis*-diols by oxidation of C=C double bonds. For example, cyclohexane diol is prepared from cyclohexene. Since these oxides are very volatile and poisonous, they should be handled very carefully.

1-dimensional Chain-like Oxide

Mercury oxide, HgO, is a red crystallline compound that is formed when mercury nitrate is heated in air. HgO has an infinite zigzag structure. Chromium trioxide, CrO_3, is a red crystalline compound with a low melting point (197°C) and its structure is composed of CrO_4 tetrahedra connected in one dimension. The acidity and oxidizing power of chromium trioxide are very high. It is used as an oxidation reagent in organic chemistry.

Two dimensional Stratified Oxides

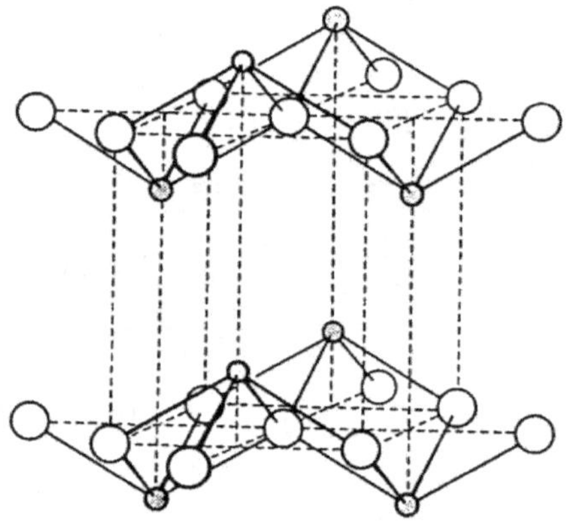

Figure 4.11: Structure of PbO

Tetragonal and blue black tin oxide, SnO, and red lead oxide, PbO, are layer compounds composed of square pyramids with the metal atom at the peak and four oxygen atoms at the bottom vertices. The structure contains metal atoms above and below the layer of oxygen atoms alternately and in parallel with the oxygen layers (Figure 4.11). Molybdenum trioxide, MoO_3, is formed by burning the metal in oxygen and shows weak oxidizing power in aqueous alkaline solutions. It has a 2-dimensional lamellar structure in which the chains of edge-sharing octahedra MoO_6 are corner-linked.

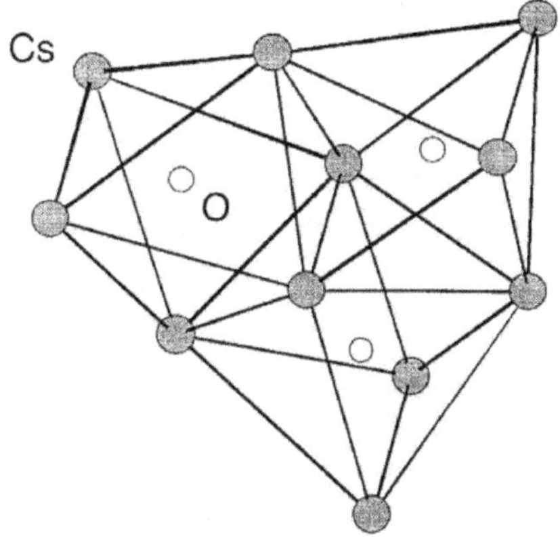

Figure 4.12: Structure of $Cs_{11}O_3$

3-dimensional Oxides

Alkali metal oxides, M_2O (M is Li, Na, K, and Rb), have the antifluorite structure (refer to Section 2.2 (e)), and Cs_2O is the anti-$CdCl_2$ lamellar structure (refer to Section 4.5 (d)). M_2O forms together with peroxide M_2O_2 when an alkali metal burns in air, but M_2O becomes the main product if the amount of oxygen is less than stoichiometric. Alternatively, M_2O is obtained by the pyrolysis of M_2O_2 after complete oxidation of the metal. Peroxide M_2O_2 (M is Li, Na, K, Rb, and Cs) can be regarded also as the salts of dibasic acid H_2O_2. Na_2O_2 is used industrially as a bleaching agent. Superoxide MO_2 (M is K, Rb, and Cs) contains paramagnetic ion O_2^-, and is stabilized by the large alkali metal cation. If there is a deficit of oxygen during the oxidation reactions of alkali metals, suboxides like Rb_9O_2 or $Cs_{11}O_3$ form. These suboxides exhibit metallic properties and have interesting cluster structures (Figure 4.12). Many other oxides in which the ratio of an alkali metal and oxygen varies, such as M_2O_3, have also been synthesized.

MO Type Metal Oxides

Except for BeO (Wurtz type), the basic structure of Group 2 metal oxides MO is the rock salt structure. They are obtained by calcination of the metal carbonates. Their melting points are very high and all are refractory. Especially quicklime, CaO, is produced and used in large quantities. The basic structure of transition metal oxides MO (M is Ti, Zr, V, Mn, Fe, Co, Ni, Eu, Th, and U) is also the rock salt structure, but they have defect structures and the ratios of a metal and oxygen are non-stoichiometric. For example, FeO has the composition Fe_xO (x = 0.89-0.96) at 1000°C. The charge imbalance of the charge is compensated by the partial oxidation of Fe^{2+} into Fe^{3+}. NbO has a defective rock salt-type structure where only three NbO units are contained in a unit cell.

MO_2 Type Metal Oxides

The dioxides of Sn, Pb, and other transition metals with small ionic radii take rutile-type structures (Figure 4.13), and the dioxides of lanthanide and actinide metals with large ionic radii take fluorite-type structures.

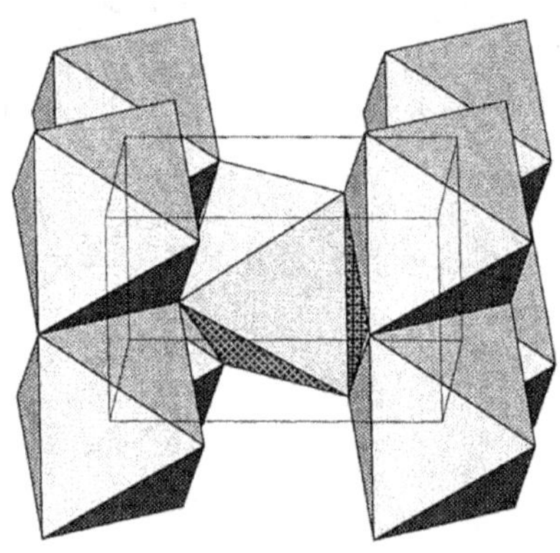

Figure 4.13: Structure of Rutile

Rutile is one of the three structure types of TiO_2, and is the most important compound used in the manufacture of the white pigments. Rutile has also been extensively studied as a water photolysis catalyst. As shown in Figure 4.13, the rutile-type structure has TiO_6 octahedra connected by the edges and sharing corners. It can be regarded as a deformed hcp array of oxygen atoms in which one half of the octahedral cavities are occupied by titanium atoms. In the normal rutile-type structure, the distance between adjacent M atoms in the edge-sharing octahedra is equal, but some rutile-type metal oxides that exhibit semiconductivity have unequal M-M-M distances. CrO_2, RuO_2, OsO_2, and IrO_2 show equal M-M distances and exhibit metallic conductivity.

Manganese dioxide, MnO_2, tends to have a non-stoichiometric metal-oxygen ratio when prepared by the reaction of manganese nitrate and air, although the reaction of manganese with oxygen gives almost stoichiometric MnO_2 with a rutile structure. The following reaction of manganese dioxide with hydrochloric acid is useful for generating chlorine in a laboratory.

$$MnO_2 + 4\,HCl \longrightarrow MnCl_2 + Cl_2 + 2\,H_2O$$

Zirconium dioxide, ZrO_2, has a very high melting-point (2700°C), and is resistant to acids and bases. It is also a hard material and used for crucibles or firebricks. However, since pure zirconium dioxide undergoes phase transitions at 1100 °C and 2300°C that result in it breaking up, solid solutions with CaO or MgO are used as fireproof materials. This is called **stabilized zirconia.**

M_2O_3-type Oxides

The most important structure of the oxides of this composition is the **corundum structure** (Al, Ga, Ti, V, Cr, Fe, and Rh). In the corundum structure, 2/3 of the octahedral cavities in the hcp array of oxygen atoms are occupied by M^{3+}. Of the two forms of alumina, Al_2O_3, α alumina and γ alumina, α alumina takes the corundum structure and is very hard. It is unreactive to water or acids. Alumina is the principal component of jewelry, such as ruby and sapphire. Moreover, various **fine ceramics** (functional porcelain materials) utilizing the properties of α-alumina have been developed. On the other hand, γ alumina has a defective spinel-type structure, and it adsorbs water and dissolves in acids, and is the basic component of activated alumina. It has many chemical uses including as a catalyst, a catalyst support, and in chromatography.

MO_3 Type Oxides

Rhenium and tungsten oxides are important compounds with this composition. **Rhenium trioxide,** ReO_3, is a dark red compound prepared from rhenium and oxygen that has a metallic luster and conductivity. ReO_3 has a three-dimensional and very orderly array of ReO_6 regular, corner-sharing octahedra (Figure 4.14).

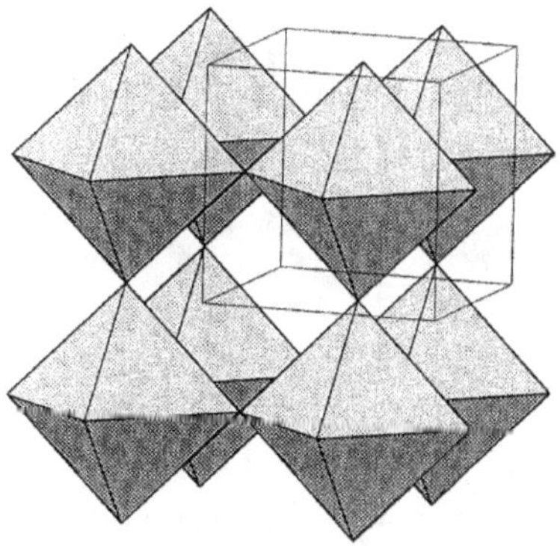

Figure 4.14: Structure of ReO_3

Tungsten trioxide, WO_3, is the only oxide that shows various phase transitions near room temperature and at least seven polymorphs are known. These polymorphs have the ReO_3-type three-dimensional structure with corner-sharing WO_6 octahedra. When these compounds are heated in a vacuum or with powdered tungsten, reduction takes place and many oxides with complicated compositions ($W_{18}O_{49}$, $W_{20}O_{58}$, *etc.*) are formed. Similar molybdenum oxides are known and they had been regarded as non-stoichiometric compounds before A. Magneli found that they were in fact stoichiometric compounds.

Mixed Metal Oxides

Spinel, $MgAl_2O_4$, has a structure in which Mg^{2+} occupy 1/8 of the tetrahedral cavities and Al^{3+} 1/2 of the octahedral cavities of a ccp array of oxygen atoms (Figure 4.15) Among the oxides of composition $A^{2+}B_2^{3+}O_4$ (A^{2+} are Mg, Cr, Mn, Fe, Co, Ni, Cu, Zn, Cd, Sn, and B^{3+} are Al, Ga, In, Ti, V, Cr, Mn, Fe, Co, Ni, and Rh) , those in which the tetrahedral holes are occupied by A^{2+} or B^{3+} are called **normal spinels** or **inverse spinels**, respectively. Spinel itself has a normal spinel-type structure, and $MgFe_2O_4$ and Fe_3O_4 have inverse spinel-type structures. Crystal field stabilization energies (differ depending on whether the crystal field of the oxygen atoms is a regular tetrahedron or octahedron). Therefore, when the metal component is a transition metal, the energy difference is one of the factors to determine which of A^{2+} or B^{3+} is favorable to fill the tetrahedral cavities.

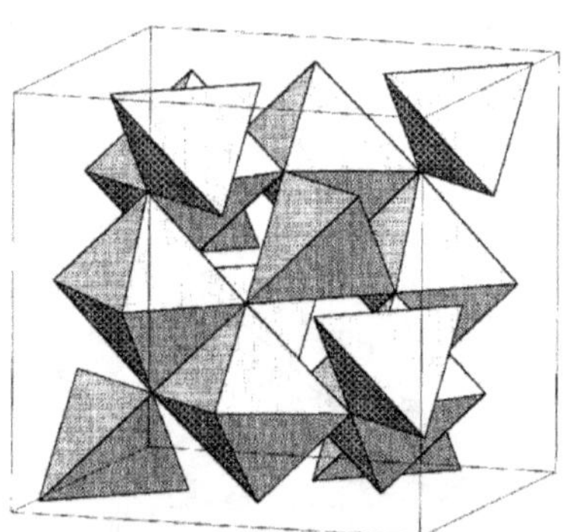

Figure 4.15: Spinel Structure

Perovskite, $CaTiO_3$, is an ABO_3 oxide (the net charge of A and B becomes 6+), and it has a structure with calcium atom at the center of TiO_3 in the ReO_3 structure (Figure 4.16). Among this kind of compounds, $BaTiO_3$, commonly called barium titanate, is especially important. This ferroelectric functional material is used in nonlinear resistance devices (varistor).

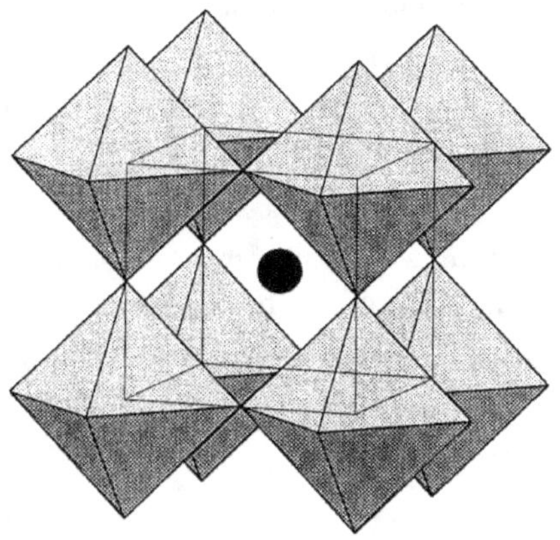

Figure 4.16: Perovskite Structure

(h) Oxides of Group 14 Elements

Although GeO_2 has a rutile-type structure, there is also a β quartz-type polymorphism. There are germanium oxides with various kinds of structures analogous to silicates and aluminosilicates. SnO_2 takes a rutile-type structure. SnO_2 is used in transparent electrodes, catalysts, and many other applications. Surface treatment with tin oxide enhances heat reflectivity of glasses. PbO_2 usually has a rutile-type structure. Lead oxide is strongly oxidizing and used for the manufacture of chemicals, and PbO_2 forms in a lead batteries.

(i) Isopolyacids, Heteropolyacids, and their Salts

There are many polyoxo acids and their salts of Mo(VI) and W (VI). V (V), V (IV), Nb (V), and Ta (V) form similar polyoxo acids although their number is limited. **Polyoxoacids** are polynuclear anions formed by polymerization of the MO_6 coordination polyhedra that share corners or edges. Those consisting only of metal, oxygen, and hydrogen atoms are called **isopolyacids** and those containing various other elements (P, Si, transition metals, *etc.*) are called **heteropolyacids**. The salts of polyacids have counter-cations such as sodium or ammonium instead of protons. The history of polyoxoacids is said to have started with J. Berzelius discovering the first polyoxoacid in 1826, with the formation of yellow precipitates when he acidified an aqueous solution containing Mo (VI) and P (V). The structures of polyoxoacids are now readily analyzed with single crystal X-ray structural analysis, ^{17}O NMR, *etc.* Because of their usefulness as industrial catalysts or for other purposes, polyoxoacids are again being studied in detail.

Keggin-structure. The **heteropolyoxo** anions expressed with the general formula $[X^{n+}M_{12}O_{40}]^{(8-n)-}$ (M = Mo, W, and X = B, Al, Si, Ge, P, As, Ti, Mn, Fe, Co, Cu, *etc*.) have the **Keggin structure,** elucidated by J. F. Keggin in 1934 using X-ray powder diffraction. For example, the structure of the tungstate ion containing silicon, in which 12 WO_6 octahedra enclose the central SiO_4 tetrahedron and four groups of three edge-shared octahedra connect to each other by corner sharing, is shown in Figure 4.17. The four oxygen atoms that coordinate to the silicon atom of the SiO_4 tetrahedra also share three WO_6 octahedra. . Therefore, the whole structure shows T_d symmetry. Although the Keggin structure is somewhat complicated, it is very symmetrical and beautiful and is the most typical structure of heteropolyoxo anions. Many other types of heteropolyoxo anions are known.

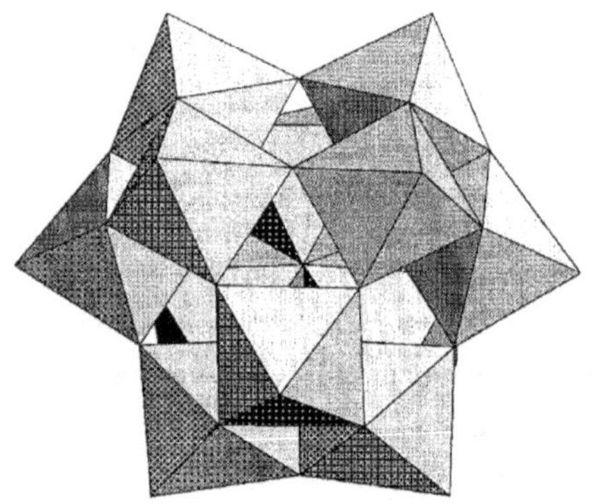

Figure 4.17: Keggin Structure

Polyoxo anions are generated by the condensation of MO_6 units by removal of H_2O when MoO_4^{2-} reacts with a proton H^+, as is shown in the following equation.

$$12\ [MoO_4]^{2-} + HPO_4^{2-} \xrightarrow{\ H^+\ } [PMo_{12}O_{40}]^{3-} + 12\ H_2O$$

Therefore, the size and form of heteropolyoxo anions in the crystal precipitation are decided by the choice of acid, concentration, temperature, or the counter cation for crystallization. A number of studies on the solution chemistry of dissolved anions have been performed.

Heteropolyoxo anions display notable oxidizing properties. As heteropolyoxo anions contain metal ions of the highest oxidation number, they are reduced even by very weak reducing agents and show mixed valence. When Keggin-type anions are reduced by one electron, they show a very deep-blue color. It has been proved that the Keggin structure is preserved at this stage and polyoxo anions absorb more electrons and several M(V) sites are generated. Thus, a heteropolyoxo anion can serve as an electron sink for many electrons, and heteropolyoxo anions exhibit photo-redox reactions.

Problem 4.4: What is the major difference in the structures of a polyacid and a solid acid?

[Answer] Although polyacids are molecules with definite molecular weights, the usual solid oxides have an infinite number of metal-oxygen bonds.

4.4.   Chalcogen and Chalcogenides

(a) Simple Substances

Sulfur, selenium, and tellurium are called chalcogens. Simple substances and compounds of oxygen and of the elements of this group in the later periods have considerably different properties. As a result of having much smaller electro negativities than oxygen, they show decreased ionicity and increased bond covalency, resulting in a smaller degree of hydrogen bonding. Because they have available d orbitals, chalcogens have increased flexibility of valence and can easily bond to more than two other atoms.

Catenation is the bonding between the same chalcogen atoms, and both simple substances and ions of chalcogens take a variety of structures.

The major isotopes of sulfur are ^{32}S (95.02% abundance), ^{33}S (0.75%), ^{34}S (4.21%), and ^{36}S (0.02%), and there are also six radioactive isotopes. Among these, ^{33}S ($I = 3/2$) can be used for NMR. Since the isotope ratio of sulfurs from different locations differs, the accuracy of the atomic weight is limited to 32.07+0.01. Because the electronegativity of sulfur ($\chi = 2.58$) is much smaller than that of oxygen ($\chi = 3.44$) and sulfur is a soft element, the ionicity in the bonds of sulfur compounds is low and hydrogen bonding is not important. Elemental sulfur has many allotropes, such as S_2, S_3, S_6, S_7, S_8, S_9, S_{10}, S_{11}, S_{12}, S_{18}, S_{20}, and S_∞, reflecting the catenation ability of sulfur atoms.

Elemental sulfur is usually a yellow solid with a melting point of 112.8°C called orthorhombic sulfur (α sulfur). Phase-transition of this polymorph produces monoclinic sulfur (β sulfur) at 95.5°C. It was established in 1935 that these are crown-like cyclic molecules (Figure 4.18). Being molecular, they dissolve well in organic solvents, such as CS_2. Not only 0-membered rings but also $S_{6\text{-}20}$ rings are known, and the helix polymer of sulfur is an infinitely annular sulfur. Diatomic molecular S_2 and triatomic molecular S_3 exist in the gaseous phase. When sulfur is heated, it liquifies and becomes a rubber-like macromolecule on cooling. The diversity of structures of catenated sulfur is also seen in the structures of the polysulfur cations or anions resulting from the redox reactions of the catenated species.

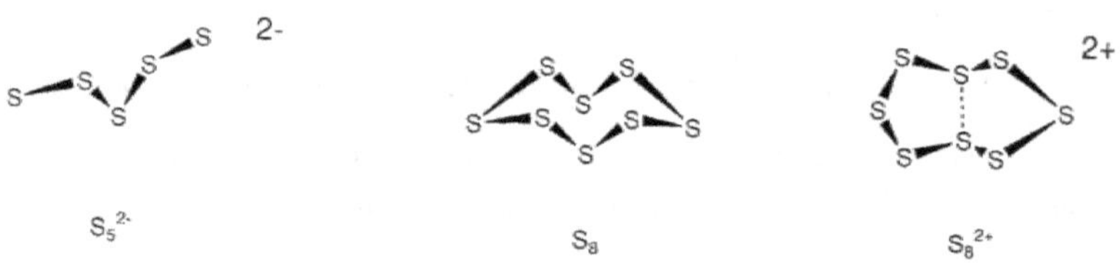

Figure 4.18: Structures of S_5^{2-}, S_8, and S_8^{2+}

Selenium is believed to have six isotopes. ^{80}Se (49.7%) is the most abundant and ^{77}Se, with nuclear spin $I = 1/2$ is useful in NMR. The accuracy of atomic weight of selenium, 78.96+0.03, is limited to two decimal places because of composition change of its isotopes. Among many allotropes of selenium, so-called red selenium is an Se_8 molecule with a crown-like structure and is soluble in CS_2. Gray metallic selenium is a polymer with a helical structure. Black selenium, which is a complicated polymer, is also abundant.

Tellurium also has eight stable isotopes and an atomic weight of 127.60+0.03.

^{130}Te (33.8%) and 128Te (31.7%) are the most abundant isotopes, and ^{125}Te and ^{123}Te with $I = 1/2$ can be used in NMR. There is only one crystalline form of tellurium, which is a spiral chain polymer that shows electric conductivity.

(b) Polyatomic Chalcogen Cations and Anions

Although it has long been recognized that solutions of chalcogen elements in sulfuric acid showed beautiful blue, red, and yellow colors, the polycationic species that give rise to these colors, S_4^{2+}, S_6^{2+}, S_6^{4+}, S_8^{2+}, S_{10}^{2+}, S_{19}^{2+}, or those of other chalcogen atoms, have been isolated by the reaction with AsF_5, *etc.* and their structures determined. For example, unlike neutral S_8, S_8^{2+} takes a cyclic structure that has a weak coupling interaction between two transannular sulfur atoms (Figure 4.18).

On the other hand, alkali metal salts Na_2S_2, K_2S_5, and alkaline earth metal salt BaS_3, a transition-metal salt $[Mo_2(S_2)_6]^{2-}$, a complex $Cp_2W(S_4)$, *etc.* of polysulfide anions S_x^{2-} (x = 1-6), in which the sulfur atoms are bonded mutually have been synthesized and their structures determined. As is evident from the fact that elemental sulfur itself forms S_8 molecules, sulfur, unlike oxygen, tends to catenate. Therefore, formation of polysulfide ions, in which many sulfur atoms are bonded, is feasible, and a series of polysulfanes H_2S_x (x = 2-8) has actually been synthesized.

(c) Metal Sulfides

Stratified disulfides, MS_2, are important in transition metal sulfides. They show two types of structures. One has a metal in a triangular prismatic coordination environment, and the other has a metal in an octahedral coordination environment.

MoS_2 is the most stable black compound among the molybdenum sulfides. L. Pauling determined the structure of MoS_2 in 1923. The structure is constructed by laminating two sulfur layers between which a molybdenum layer is intercalated (Figure 4.19). Alternatively, two sulfur layers are stacked and a molybdenum layer is inserted between them. Therefore, the coordination environment of each molybdenum is a triangular prism of sulfur atoms. Since there is no bonding interaction between sulfur layers, they can easily slide, resulting in graphite-like lubricity. MoS_2 is used as a solid lubricant added to gasoline, and also as a catalyst for hydrogenation reactions.

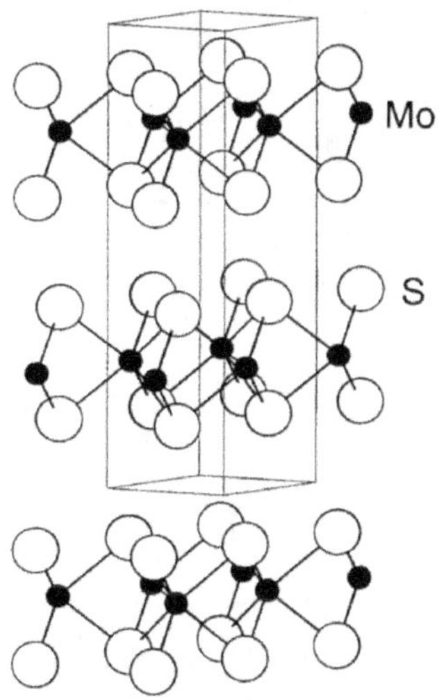

Figure 4.19: Structure of MoS_2

ZrS_2, TaS_2, *etc.* take the CdI_2type structure containing metal atoms in an octahedral coordination environment constructed by sulfur atoms.

Chevrel phase compounds. There are superconducting compounds called **Chevrel phases** which are important examples of the chalcogenide compounds of molybdenum. The general formula is described by $M_xMo_6X_8$ (M = Pb, Sn, and Cu; X = S, Se, and Te), and six molybdenum atoms form a regular octahedral cluster, and eight chalcogenide atoms cap the eight triangular faces of the cluster. The cluster units are connected 3-dimensionally (Figure 4.20). Since the cluster structure of molybdenum atoms is similar to that of molybdenum dichloride, $MoCl_2=((Mo_6Cl_8)Cl_2Cl_{4/2})$, the structural chemistry of these compounds has attracted as much attention as their physical properties.

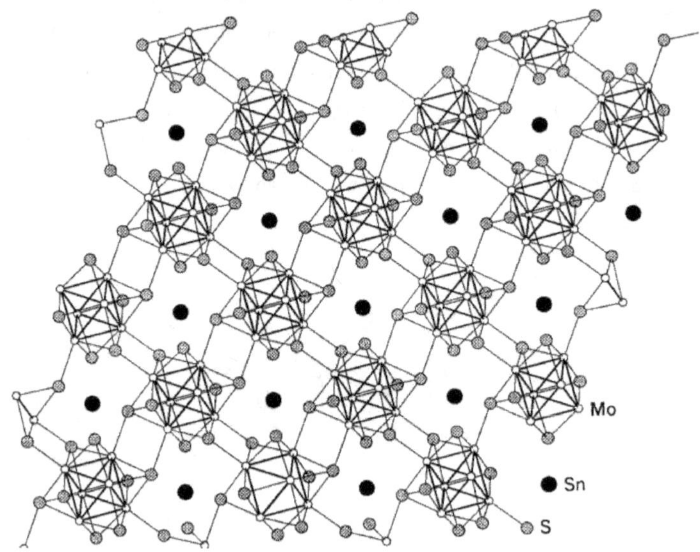

Figure 4.20: Structure of $SnMo_6S_8$

4.5. Halogens and Halides

The origin of halogen is the Greek word meaning the production of salt by direct reaction with a metal. Since their reactivity is very high, halogens are found in nature only as compounds. The basic properties of halogens are shown in Table 4.6 and Table 4.7. The electron configuration of each halogen atom is ns^2np^5, and they lack one electron from the closed-shell structure of a rare gas. Thus a halogen atom emits energy when it gains an electron. Namely, the enthalpy change of the reaction $X(g) + e^- \rightarrow X^- (g)$ is negative. Although electron affinity is defined as the energy change of gaining an electron, a positive sign is customarily used. In order to be consistent with the enthalpy change, a negative sign would be appropriate.

Table 4.6: Properties of Halogens

	Ionization energy ($kJ\,mol^{-1}$)	Electronegativity χP	Ionic radius $r(X^-)$ (pm)
F	1680.6	3.98	133
Cl	1255.7	3.16	181
Br	1142.7	2.96	196
I	1008.7	2.66	220

Table 4.7: Properties of Halogen Molecules

	Interatomic distance $r(X\text{-}X)$ (pm)	mp °C	bp °C	Color
F_2	143	-218.6	-188.1	Colorless gas
Cl_2	199	-101.0	-34.0	Yellow green gas
Br_2	228	-7.75	59.5	Dark red liquid
I_2	266	113.6	185.2	Dark violet solid

The electron affinity of chlorine (348.5 kJmol⁻¹) is the largest and fluorine (332.6 kJmol⁻¹) comes between chlorine and bromine (324.7 kJmol⁻¹). The electronegativity of fluorine is the highest of all the halogens.

Since halogens are produced as metal salts, simple substances are manufactured by electrolysis. Fluorine only takes the oxidation number -1 in its compounds, although the oxidation number of other halogens can range from -1 to +7. Astatine, At, has no stable nuclide and little is known about its chemical properties .

(a) Manufacture of Halogen

Fluorine has the highest reduction potential (E^0 = +2.87 V) and the strongest oxidizing power among the halogen molecules. It is also the most reactive nonmetallic element. Since water is oxidized by F_2 at much lower electrode potential (+1.23 V), fluorine gas cannot be manufactured by the electrolysis of aqueous solutions of fluorine compounds. Therefore, it was a long time before elemental fluorine was isolated, and F. F. H. Moisson finally succeeded in isolating it by the electrolysis of KF in liquid HF. Fluorine is still manufactured by this reaction. Chlorine, which is especially important in inorganic industrial chemistry, is manufactured together with sodium hydroxide. The basic reaction for the production of chlorine is electrolysis of an aqueous solution of NaCl using an ion exchange process. In this process, chlorine gas is generated in an anodic cell containing brine and Na^+ moves through an ion exchange membrane to the cathodic cell where it pairs with OH^- to become an aqueous solution of NaOH

Problem 4.5: Why can chlorine be manufactured by electrolysis of an aqueous solution of sodium chloride?

[Answer] Despite the higher reduction potential of chlorine (+1.36 V) than that of oxygen (+1.23 V), the reduction potential of oxygen can be raised (overvoltage) depending on the choice of electrode used for the electrolysis process

Bromine is obtained by the oxidation of Br^- with chlorine gas in saline water. Iodine is similarly produced by passing chlorine gas through saline water containing I^- ions. Since natural gas is found in Japan together with underground saline water containing I^-, Japan is one of the main countries producing iodine.

Anomalies of fluorine: Molecular fluorine compounds have very low boiling points. This is due to the difficulty of polarization as a result of the electrons being strongly drawn to the nuclei of fluorine atoms. Since the electronegativity of fluorine is highest (χ = 3.98) and electrons shift to F, resulting in the high acidity of atoms bonded to F. Because of the small ionic radius of F^-, high oxidation states are stabilized, and hence low oxidation compounds like CuF are unknown, in contrast with the compounds such as IF_7 and PtF_6.

Pseudohalogens: Since the cyanide ion CN^-, the azide ion N^{3-}, and the thiocyanate ion SCN^-, *etc.* form compounds similar to those of halide ions, they are called **pseudohalide ions**. They form psudohalogen molecules such as cyanogene $(CN)_2$, hydrogen cyanide HCN, sodium thiocyanate NaSCN, *etc.* Fine-tuning electronic and steric effects that are impossible with only halide ions make pseudohalogens useful also in transition metal complex chemistry.

Polyhalogens: Besides the usual halogen molecules, mixed halogen and polyhalogen molecules such as BrCl, IBr, ICl, ClF_3, BrF_5, IF_7 *etc* also exist. Polyhalogen anions and cations such as I_3^-, I_5^-, I_3^+, and I_5^+, are also known.

(b) Oxygen Compounds

Although many binary oxides of halogens (consisting only of halogen and oxygen) are known, most are unstable. Oxygen difluoride OF_2 is the most stable such compound. This is a very powerful fluorinating agent and can generate plutonium hexafluoride PuF_6 from plutonium metal. While oxygen chloride, Cl_2O, is used for bleaching pulp and water treatment, it is generated *in situ* from ClO_3^-, since it is unstable.

Hypochlorous acid, $HClO$, chlorous acid, $HClO_2$, chloric acid, $HClO_3$, and perchloric acid, $HClO_4$ are oxoacids of chlorine and especially perchloric acid is a strong oxidizing agent as well as being a strong acid. Although analogous acids and ions of other halogens had been known for many years, BrO_4^- was synthesized as late as 1968. Once it was prepared it turned out to be no less stable than ClO_4^- or IO_4^-, causing some to wonder why it had not been synthesized before. Although ClO_4^- is often used for crystallizing transition metal complexes, it is explosive and should be handled very carefully.

(c) Halides of Nonmetals

Halides of almost all nonmetals are known, including fluorides of even the inert gases krypton, Kr, and xenon, Xe. Although fluorides are interesting for their own unique characters, halides are generally very important as starting compounds for various compounds of nonmetals by replacing halogens in inorganic syntheses (Table 4.8).

Table 4.8: Typical Chlorides and Fluorides of Main Group Elements

	1	2	12	13	14	15	16	17	18
2	LiCl	$BeCl_2$		BF_3	CCl_4	NF_3	OF_2		
3	NaCl	$MgCl_2$		$AlCl_3$	$SiCl_4$	PCl_3 PCl_5	S_2Cl_2 SF_6	ClF_3 ClF_5	
4	KCl	$CaCl_2$	$ZnCl_2$	$GaCl_3$	GeF_2 $GeCl_4$	$AsCl_3$ AsF_5	Se_2Cl_2 SeF_5	BrF_3 BrF_5	KrF_2
5	RbCl	$SrCl_2$	$CdCl_2$	InCl $InCl_3$	$SnCl_2$ $SnCl_4$	$SbCl_3$ SbF_5	Te_4Cl_{16} TeF_6	IF_5 IF_7	XeF_2 XeF_6
6	CsCl	$BaCl_2$	Hg_2Cl_2 $HgCl_2$	TlCl $TlCl_3$	$PbCl_2$ $PbCl_4$	$BiCl_3$ BiF_5			

Boron trifluoride, BF_3, is a colorless gas (mp -127°C and bp -100°C) that has an irritating odor and is poisonous. It is widely used as an industrial catalyst for Friedel-Crafts type reactions. It is also used as a catalyst for cationic polymerization. It exists in the gaseous phase as a triangular monomeric molecule, and forms Lewis base adducts with ammonia, amines, ethers, phosphines, *etc.* because of its strong Lewis acidity. Diethylether adduct, $(C_2H_5)_2O{:}BF_3$, is a distillable liquid and is used as a common reagent. It is a starting compound for the preparation of diborane, B_2H_6. Tetrafluoroborate, BF_4^-, is a tetrahedral anion formed as an adduct of BF_3 with a base F^-. Alkali metal salts, a silver salt and $NOBF_4$ as well as the free acid HBF_4 contain this anion. Since its coordination ability is very weak, it is used in the crystallization of cationic complexes of transition metals as a counter anion like ClO_4^-. $AgBF_4$ and $NOBF_4$ are also useful for 1-electron oxidation of complexes.

Tetrachlorosilane, $SiCl_4$, is a colorless liquid (mp -70°C and bp 57.6°C). It is a regular tetrahedral molecule, and reacts violently with water forming silicic acid and hydrochloric acid. It is useful as a raw material for the production of pure silicon, organic silicon compounds, and silicones.

Phosphorus trifluoride, PF_3, is a colorless, odorless, and deadly poisonous gas (mp -151.5 °C and bp -101.8°C). This is a triangular pyramidal molecule. Because it is as electron-attracting as CO, it acts as a ligand forming metal complexes analogous to metal carbonyls.

Phosphorus pentafluoride, PF_5, is a colorless gas (mp -93.7°C and bp -84.5°C). It is a triangular bipyramidal molecule and should have two distinct kinds of fluorine atoms. These fluorines exchange positions so rapidly that they are indistinguishable by ^{19}F NMR. It was the first compound with which the famous Berry's pseudorotation was discovered as an exchange mechanism for axial and equatorial fluorine atoms. The hexafluorophosphate ion, PF_6^-, as well as BF_4^- is often used as a counter anion for cationic transition metal complexes. $LiPF_6$ and R_4NPF_6 can be used as supporting electrolytes for electrochemical measurements.

Phosphorus trichloride, PCl_3, is a colorless fuming liquid (mp -112°C and bp 75.5°C). It is a triangular pyramidal molecule and hydrolyzes violently. It is soluble in organic solvents. It is used in large quantities as a raw material for the production of organic phosphorus compounds.

Phosphorus pentachloride, PCl_5, is a colorless crystalline substance (sublimes but decomposes at 160°C) It is a triangular bipyramidal molecule in the gaseous phase, but it exists as an ionic crystal $[PCl_4]^+[PCl_6]^-$ in the solid phase. Although it reacts violently with water and becomes phosphoric acid and hydrochloric acid, it dissolves in carbon disulfide and carbon tetrachloride. It is useful for chlorination of organic compounds.

Arsenic pentafluoride, AsF_5, is a colorless gas (mp -79.8°C and bp -52.9°C). It is a triangular bipyramidal molecule. Although it hydrolyzes, it is soluble in organic solvents. As it is a strong electron acceptor, it can form electron donor-acceptor complexes with electron donors.

Sulfur hexafluoride, SF_6, is a colorless and odorless gas (mp -50.8°C and sublimation point -63.8 °C) It is a hexacoordinate octahedral molecule. It is chemically very stable and hardly soluble in water. Because of its excellent heat-resisting property, incombustibility, and corrosion resistance, it is used as a high voltage insulator.

Sulfur chloride, S_2Cl_2, is an orange liquid (mp -80°C and bp 138°C). It has a similar structure to hydrogen peroxide. It is readily soluble in organic solvents. It is important as an industrial inorganic compound, and is used in large quantities for the vulcanization of rubber etc.

(d) Metal Halides

Many metal halides are made by the combination of about 80 metallic elements and four halogens (Table 4.8, Table 4.9). Since there are more than one oxidation state especially in transition metals, several kinds of halides are known for each transition metal. These halides are most important as starting materials of the preparation of metal compounds, and the inorganic chemistry of metal compounds depends on metal halides. There is molecular, 1-dimensional chain, 2-dimensional layer, and 3-dimensional halides but few of them are molecular in crystalline states. It should be noted that the anhydrous transition metal halides are usually solid compounds and hydrates are coordination compounds with water ligands. As the dimensionality of structures is one of the most interesting facets of structural or synthetic chemistry, typical halides are described in order of their dimensionality.

Table 4.9: Typical Chlorides and Fluorides of Transition Metals

Oxidation Number	3	4	5	6	7	8	9	10	11
+1	$ScCl$ YCl $LaCl$	$ZrCl$ $HfCl$							$CuCl$ $AgCl$ $AuCl$
+2		$TiCl_2$	VCl_2	$CrCl_2$ $MoCl_2$ WCl_2	$MnCl_2$	$FeCl_2$ $RuCl_2$	$CoCl_2$	$NiCl_2$ $PdCl_2$ $PtCl_2$	$CuCl_2$
+3	ScF_3 YCl_3 LaF_3	$TiCl_3$ $ZrCl_3$	VCl_3	$CrCl_3$ $MoCl_3$ WCl_3	$ReCl_3$	$FeCl_3$ $RuCl_3$ $OsCl_3$	CoF_3 $RhCl_3$ $IrCl_3$		$AuCl_3$
+4		$TiCl_4$ $ZrCl_4$ $HfCl_4$	VCl_4 $NbCl_4$ $TaCl_4$	CrF_4 $MoCl_4$ WCl_4	$ReCl_4$			$PtCl_4$	
+5			VF_5 $NbCl_5$ $TaCl_5$	CrF_5 $MoCl_5$ WCl_5	$ReCl_5$	OsF_5	IrF_5	PtF_5	
+6				WCl_6	ReF_6	OsF_6	IrF_6	PtF_6	
+7					ReF_7	OsF_7			

Molecular Halides

Mercury(II) chloride, $HgCl_2$. It is a colorless crystal soluble in water and ethanol. It is a straight, three-atomic molecule in the free state. However, in addition to two chlorine atoms bonded to mercury, four additional chlorine atoms of adjacent molecules occupy coordination sites and the mercury is almost hexacoordinate in the crystalline state. The compound is very toxic and used for preserving wood, *etc.*

Aluminum trichloride, $AlCl_3$. A colorless crystal (mp 190°C (2.5 atm) and bp183°C) that sublimes when heated. It is soluble in ethanol and ether. It is a Lewis acid and forms adducts with various bases. It is a molecule consisting of the dimer of tetracoordinate aluminium with chlorine bridges in the liquid and gaseous phases (Figure 4.21), and takes a lamellar structure when crystalline. It is used as a Lewis acid catalyst of Friedel-Crafts reactions, *etc.*

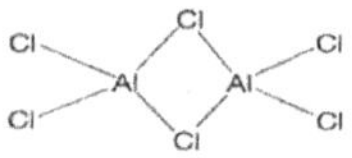

Figure 4.21: Structure of Aluminum Chloride

Tin (IV) chloride, $SnCl_4$. A colorless liquid (mp -33°C and bp 114°C). In the gaseous state, it is a tetrahedral molecule.

Titanium(IV) chloride, $TiCl_4$. A colorless liquid (mp -25°C and bp 136.4°C). The gaseous molecule is a tetrahedron similar to tin(IV) chloride. It is used as a component of the Ziegler Natta catalyst.

Chain-like Halides

Gold (I) iodide, AuI. Yellow white solid. Two iodines coordinate to gold, and the compound has a zigzag 1-dimensional chain structure.

Beryllium chloride, $BeCl_2$. A colorless crystal (mp 405°C and bp 520°C). It is deliquescent and soluble in water and ethanol. The tetra-coordinated beryllium forms a 1-dimensional chain via chlorine bridges (Figure 4.22). In the gaseous phase, it is a straight two-coordinate molecule. It is a Lewis acid and is used as a catalyst for Friedel-Crafts reactions.

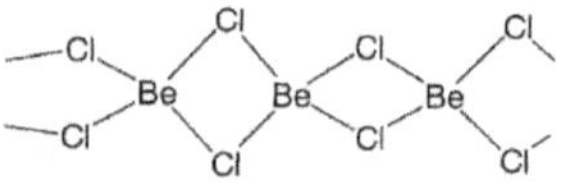

Figure 4.22: Structure of Beryllium Chloride

Palladium chloride, $PdCl_2$. A dark red solid. In the α type, the four-coordinate palladium forms a 1-dimensional chain with double bridges of chlorines . The dehydrate is deliquescent and soluble in water, ethanol, acetone, *etc.* When it is dissolved in hydrochloric acid, it becomes four-coordinate square-planar $[PdCl_4]^{2-}$. It is used as the catalyst for the Wacker process, which is an olefin oxidation process, or in various catalysts for organic syntheses.

Zirconium tetrachloride, (IV) $ZrCl_4$. A colorless crystal (it sublimes above 331 ºC). The zirconium is octahedrally coordinated and forms a zigzag chain via chlorine bridges (Figure 4.23). It is hygroscopic and soluble in water, ethanol, *etc.* It is used as a Friedel-Crafts catalyst and as a component of olefin polymerization catalysts.

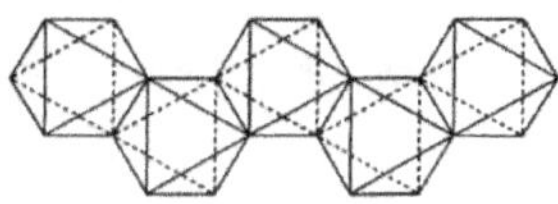

Figure 4.23: Structure of Zirconium Tetrachloride

Stratified Halides

Cadmium iodide, CdI_2. A colorless crystal (mp 388ºC and bp 787ºC). It has a cadmium iodide structure where the layers of edge-shared CdI_6 octahedral units are stratified (Figure 4.24). In the gaseous phase, it comprises straight three atomic molecules. It dissolves in water, ethanol, acetone, *etc.*

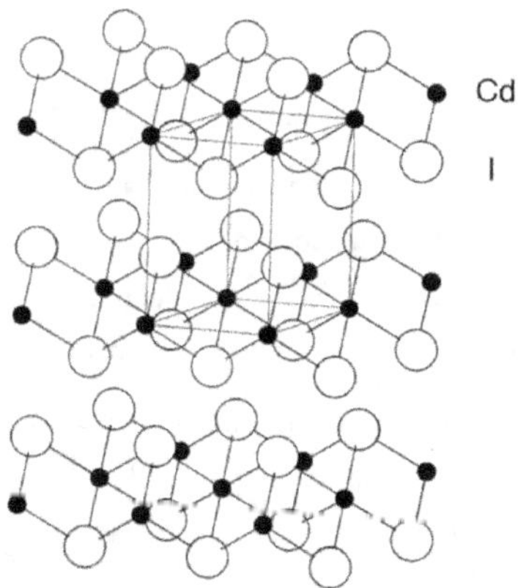

Figure 4.24: Layers of Cadmium and Iodine in the Cadmium Iodide Structure

Cobalt(II) chloride, $CoCl_2$. Blue crystals (mp 735°C and bp 1049°C). It has the cadmium chloride structure. It is hygroscopic and becomes light red when water is absorbed. It is soluble also in ethanol and acetone. The hexahydrate is red and is a coordination compound in which water molecules are ligands.

Iron (II)chloride, $FeCl_2$. Greenish yellow crystals (mp 670-674°C). It has the cadmium chloride structure, and is soluble in water and ethanol. The hydrates, which are coordinated by various numbers (6,4,2) of water molecules, are precipitated from aqueous solutions of hydrochloric acid.

Iron(III) chloride, $FeCl_3$. Dark brown crystals (mp 306°C and sublimes). It has a lamellar structure in which iron is octahedrally surrounded by six chlorine ligands. In the gaseous phase, it has a dimeric structure bridged by chlorine atoms similar to that of aluminum chloride.

3-dimensional Structure Halides

Sodium chloride, NaCl. A colorless crystal (mp 801°C and bp 1413°C). It is the original rock salt-type structure. In the gaseous phase, this is a two-atom molecule. Although it is soluble in glycerol as well as water, it hardly dissolves in ethanol. Large single crystals are used as prisms for infrared spectrometers.

Cesium chloride, CsCl. A colorless crystal (mp 645°C, bp 1300°C). Although it has the cesium chloride type structure, it changes to the rock salt structure at 445 °C. In the gaseous phase, it is a two-atom molecule.

Copper(I) chloride, CuCl. A colorless crystal (mp 430°C and bp 1490°C) It has the zinc blende structure and four chlorines tetrahedrally coordinate to copper.

Calcium chloride, $CaCl_2$. A colorless crystal (mp 772°C and bp above 1600°C). It has a deformed rutile-type structure and calcium is octahedrally surrounded by six chlorines. It is soluble in water, ethanol, and acetone. It is deliquescent and used as a desiccant. Hydrates in which 1, 2, 4, or 6 water molecules are coordinated are known.

Calcium fluoride, CaF_2. A colorless crystal (mp 1418°C and bp 2500°C). It has the fluorite type structure. It is the most important raw material for fluorine compounds. Good quality crystals are used also as spectrometer prisms and in photographic lenses.

Chromium(II) chloride, $CrCl_2$. A colorless crystal (mp 820°C and sublimes). It has a deformed rutile-type structure. It dissolves well in water giving a blue solution.

Chromium(III) chloride, $CrCl_3$. Purplish red crystal (mp 1150°C and decomposes at 1300 °C). Cr^{3+} occupies two thirds of the octahedral cavities in every other layer of Cl^-ions, which are hexagonally close-packed. It is insoluble in water, ethanol, and acetone.

Problem 4.6 Why do solid metal halides dissolve in water?

[Answer] It is because water reacts with halides breaking the halogen bridges in the solid structures and coordinates to the resultant molecular complexes.

4.6.   Rare Gases and their Compounds

(a) Rare Gases

In the 18th century, H. Cavendish discovered an inert component in air. In 1868, a line was discovered in the spectrum of sunlight that could not be identified and it was suggested to be due to a new element, helium. Based on these facts, at the end of the 19th century W. Ramsay isolated He, Ne, Ar, Kr, and Xe and by studying their properties demonstrated that they were new elements . In spite of the nearly 1% content of argon Ar in air, the element had not been isolated until then and rare gases were completely lacking in Mendeleev's periodic table. The Nobel prize was awarded to Ramsay in 1904 for his achievement.

Rare gases are located next to the halogen group in the periodic table. Since rare gas elements have closed-shell electronic configurations, they lack reactivity and their compounds were unknown. Consequently, they were also called inert gases. However, after the discovery of rare gas compounds, it was considered more suitable to call these elements "noble gases", as is mentioned in the following chapter.

Although the abundance of helium in the universe is next to that of hydrogen, it is very rare on the Earth because it is lighter than air. Helium originated from solar nuclear reactions and was locked up in the earth's crust. It is extracted as a by-product of natural gas from specific areas (especially in North America). Since helium has the lowest boiling point (4.2 K) of all the substances, it is important for low-temperature science and superconductivity engineering. Moreover, its lightness is utilized in airships *etc*. Since argon is separated in large quantities when nitrogen and oxygen are produced from liquid air, it is widely used in metallurgy, and in industries and laboratories that require an oxygen-free environment.

(b) Rare Gas Compounds

Xenon, Xe, reacts with elements with the largest electronegativities, such as fluorine, oxygen, and chlorine and with the compounds containing these elements, like platinum fluoride, PtF_6. Although the first xenon compound was reported (1962) as $XePtF_6$, the discoverer, N. Bartlett, later corrected that it was not a pure compound but a mixture of $Xe[PtF_6]_x$ (x= 1-2). If this is mixed with fluorine gas and excited with heat or light, fluorides XeF_2, XeF_4, and XeF_6 are generated. XeF_2 has chain-like, XeF_4 square, and XeF_6 distorted octahedral structures. Although preparation of these compounds is comparatively simple, it is not easy to isolate pure compounds, especially XeF_4.

Hydrolysis of the fluorides forms oxides. XeO_3 is a very explosive compound. Although it is stable in aqueous solution, these solutions are very oxidizing. Tetroxide, XeO_4, is the most volatile xenon compound. $M[XeF_8]$ (M is Rb and Cs) are very stable and do not decompose even when heated at 400°C. Thus, xenon forms divalent to octavalent compounds. Fluorides can also be used as fluorinating reagents.

Although it is known that krypton and radon also form compounds, the compounds of krypton and radon are rarely studied as both their instability and their radioactivity make their handling problematic.

Discovery of Rare Gas Compounds

H. Bartlett studied the properties of platinum fluoride PtF_6 in the 1960s, and synthesized O_2PtF_6. It was an epoch-making discovery in inorganic chemistry when analogous experiments on xenon, which has almost equal ionization energy (1170 kJmol[-1]) to that of O_2 (1180 kJmol[-1]), resulted in the dramatic discovery of $XePtF_6$.

Rare gas compounds had not been prepared before this report, but various attempts were made immediately after the discovery of rare gases. W. Ramsay isolated rare gases and added a new group to the periodic table at the end of the 19th century. Already in 1894, F. F. H. Moisson, who is famous for the isolation of F_2, reacted a 100 cm[3] argon offered by Ramsay with fluorine gas under an electric discharge but failed to prepare an argon fluoride. At the beginning of this century, A. von Antoropoff reported the synthesis of a krypton compound $KrCl_2$, but later he concluded that it was a mistake.

L. Pauling also foresaw the existence of KrF_6, XeF_6, and H_4XeO_6, and anticipated their synthesis. In 1932, a post doctoral research fellow, A. L. Kaye, in the laboratory of D. M. L. Yost of Caltech, where Pauling was a member of faculty, attempted to prepare rare gas compounds. Despite elaborate preparations and eager experiments, attempts to prepare xenon compounds by discharging electricity through a mixed gas of xenon, fluorine, or chlorine were unsuccessful. It is said that Pauling no longer showed interest in rare gas compounds after this failure.

Although R. Hoppe of Germany predicted using theoretical considerations that the existence of XeF_2 and XeF_4 was highly likely in advance of the discovery of Bartlett, he prepared these compounds only after knowing of Bartlett's discovery. Once it is proved that a compound of a certain kind is stable, analogous compounds are prepared one after another. This has also been common in synthetic chemistry of the later period, showing the importance of the first discovery.

4.7. Reaction and Physical Properties

Organic synthesis using complexes and organometallic compounds, homogeneous catalysis, bioinorganic chemistry to elucidate biological reactions in which metals participate, and studying solid state properties such as solid state catalysis, conductivity, magnetism, optical properties are all important fields of applied inorganic chemistry. Basic inorganic chemistry, although previously less developed compared with organic chemistry, is now making fast progress and covers all elements. Construction of general theories of bonding, structure, and reaction covering molecules and solids is a major problem for the near future.

Catalytic Reactions

Catalysts reduce the activation energy of reactions and enhance the rate of specific reactions. Therefore they are crucially important in chemical industry, exhaust gas treatment and other chemical reactions. While the chemical essence of **catalysis** is obscure, practical catalysts have been developed based on the accumulation of empirical knowledge. However, while gradually we have come to understand the mechanisms of homogeneous catalysis through the development of inorganic chemistry, our understanding of surface reactions in solid catalysts is also deepening.

(a) Homogeneous Catalysis

The chemistry of catalysts that are soluble in solvents has developed remarkably since the epoch-making discovery (1965) of the **Wilkinson catalyst,** $[RhCl(PPh_3)_3]$. This complex is a purplish red compound which forms by heating $RhCl_3.3H_2O$ and PPh_3 under reflux in ethanol. When dissolved in an organic solvent, this complex is an excellent catalyst for **hydrogenation** of unsaturated hydrocarbons by H_2 at ambient temperatures and pressures to form saturated hydrocarbons, and **hydroformylation** reactions of olefins with H_2 and CO to form aldehydes.

In the past, the mechanism of catalytic reactions were generally not very clear. Before the Wilkinson catalyst, the **Reppe process,** which oligomerize aceylenes or the **Ziegler-Natta catalysts** that polymerize olefins and dienes, had been discovered and detailed studies on

homogeneous catalysis had been conducted from the viewpoint of the chemistry of complexes. Consequently, catalytic reactions are now established as a cycle of a combination of a few elementary steps that occur on the metals of catalyst complexes.

Coordination and dissociation There must be a process in which reactants such as olefins are activated and react with other reactants after being coordinated to the central metal of a complex, and they dissociate from the metal as products.

Oxidative addition Oxidative addition is one among a few key elementary reactions of metal complexes. This is a reaction of such compounds as alkali metal halides, RX, acids, HX or dihydrogen, H_2, to the metal in a complex which then dissociate into R and X, H and X, or H and H, which are bonded to the metal as two fragment anions. If other ligands on the start complex are not removed, the coordination number increases by two. As alkyl, halogen, and hydride ligands are more electronegative than the central metal, they are regarded as formally anionic ligands after coordination. Therefore, the oxidation number of the central metal increases after an addition reaction. As it is an addition reaction accompanied by oxidation of the central metal, it is called oxidative addition.

For example, in the addition reaction of an alkyl halide to a tetra-coordinate iridium(I) complex $[IrCl(CO)(PPh_3)_2]$,

$$[Ir^{I}Cl(CO)(PPh_3)_2] + RI \longrightarrow [Ir^{III}(Cl)(I)(R)(CO)(PPh_3)_2]$$

iridium becomes hexa-coordinate and undergoes two-electron oxidation from +1 to +3. Since a neutral RI molecule is added, there must be no change in the charge of the whole complex, and if an alkyl and iodine are anions, the oxidation number of the central metal should increase by 2. Similar change occur when two hydride ligands are formed as the result of the addition of dihydrogen.

The reverse reaction is called **reductive elimination**. Both oxidative and reductive reactions are very important as elementary steps in the mechanism of homogeneous catalysis involving hydrocarbons and dihydrogen.

Insertion reaction In the reaction of an alkyl or hydride ligand to shift to a carbonyl or olefin ligand coexisting on the central metal, the resultant complex appears as if a carbonyl or an olefin has inserted between the M-R or M-H bond. This is called an insertion reaction.

Reaction of a coordinated ligand This is the process in which a coordinated reactant reacts to form a product. By coordinating to a metal, the reactants take geometrically and electronically suitable conformations. It is the basis of catalyst design to control these reaction conditions.

Since a reaction is repeated while the complex used as a catalyst remains unchanged by forming a cycle of reactions, the reactants/complex ratio is very small, co-inciding with the definition of a catalyst. The catalytic cycle in hydrogenation of ethylene is illustrated

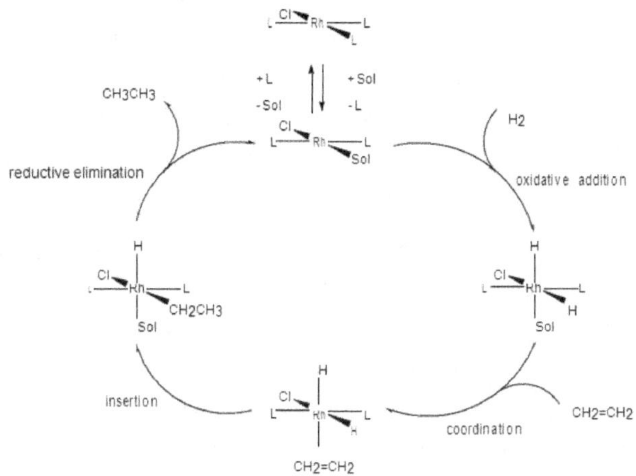

Figure 4.25: Catalytic Cycle of Ethylene Hydrogenation by the Wilkinson Catalyst. L is a ligand and Sol is a solvent molecule

If the triphenylphosphine ligand $P(C_6H_5)_3$ in the Wilkinson catalyst is replaced by an optical active phosphine, **asymmetric hydrogenation** is realized. Asymmetrical catalysis equivalent to enzyme reactions have been developed by skillful design of asymetrical ligands. In particular, the asymmetric induction of binaphtyldiphosphine (BINAP) has attracted attention.

(b) Solid state Catalysis

A solid catalyst is also called a **heterogeneous catalyst,** and promotes the reaction of reactants in gaseous or liquid phases in contact with a solid material. Since adsorption of reactants on the catalyst surface is the initial step, a large surface area is required for good efficiency of catalysis. Polyphase systems, which carry active catalysts on materials such as zeolites with small pores of molecular sizes, and gamma alumina and silica gel with large surface area, are often used.

Previously, solid state catalysis was explained as arising from a mysterious activation of reactants due to adsorption, but it has become increasingly clear that catalysis is ascribable to surface chemical reactions. Namely, the action of solid state catalysts depends on activation of reactants by surface acids or bases, and by coordination to the metal surface. It is possible to observe these interactions using various spectroscopies (infrared spectroscopy, EXAFS (extended X-ray absorption fine structure), electronic spectra), electron microscopy, or STM (scanning tunnelling microscopy).

Since mechanisms of homogeneous catalysis have been clarified considerably, solid surface reactions can also be analyzed by introducing concepts such as "surface complexes" or "surface organometallic compounds". However, unlike homogeneous catalysis, in which only one or a few metal centers participate, many active sites are involved in solid state catalysis. Since surface homogeneity and reproducibility are difficult to maintain, major parts of reaction mechanisms are obscure even for such simple reactions as ammonia synthesis.

During the direct production of ammonia from dinitrogen and dihydrogen, reactions occur using iron catalysts containing alkali metal or alkaline earth metal oxides as activators at high temperatures (about 450°C) and under high pressures (about 270 atm). Prior to the epoch-making discovery of this process by F. Haber (1909), all nitrogen compounds came from natural resources. The realization of this discovery has had an immeasurable influence upon chemical industries, as ammonia is indispensable to the manufacture of fertilizers, gunpowder and other inorganic compounds containing nitrogen. In recognition of this, a Nobel Prize was awarded to F. Haber for this invention (1918). A huge volume of research on the elucidation of the reaction mechanism of ammonia synthesis has been performed up until the present, because the reaction of dinitrogen and dihydrogen on iron catalysts is a good model of solid state catalysis.

Bioinorganic Chemistry

Many biological reactions are known to involve metal ions. There are also metals recognized as essential elements, although their roles in living organisms are not clear. Bioinorganic chemistry, the study of the functions of metals in biological systems using the knowledge and methods of inorganic chemistry, has progressed remarkably in recent years.

The following list shows typical bioactive substances containing metals.

(1) Electron carriers. Fe : cytochrome, iron-sulfur protein. Cu : blue copper protein.

(2) Metal storage compound. Fe : ferritin, transferrin. Zn : metallothionein.

(3) Oxygen transportation agent. Fe: hemoglobin, myoglobin. Cu: hemocyanin.

(4) Photosynthesis. Mg: chlorophyll.

(5) Hydrolase. Zn: carboxypeptidase. Mg: aminopeptidase.

(6) Oxidoreductase. Fe: oxygenase, hydrogenase. Fe, Mo: nitrogenase.

(7) Isomerase. Fe: aconitase. Co: vitamin B_{12} coenzyme.

The basis of chemical reactions of metalloenzymes are

(1) Coordinative activation (coordination form, electronic donating, steric effect),

(2) Redox (metal oxidation state),

(3) Information communication, and, in many cases, reaction environments are regulated by biopolymers such as proteins, and selective reactions are performed.

Examples of actions of metals other than by metalloenzymes include

(1) Mg: MgATP energy transfer

(2) Na/K ion pumping

(3) Ca: transfer of hormone functions, muscle contraction, nerve transfer, blood coagulation, are some of the important roles of metals.

(a) Oxidation

Oxidation reactions in living systems are fundamental to life, and many studies of these systems have been performed. In particular, the mechanisms of oxygen gas transportation by hemoglobin and mono-oxygen oxidation by the iron porphyrin compounds named P-450 have been studied at length. Oxygen gas transportation, which has been studied for many years, is described below. Iron porphyrins hemoglobin and myoglobin and the copper compound hemocyanin are involved in the transportation of oxygen gas in air to cells in living organisms. The basis of this function is reversible bonding and dissociation of dioxygen to iron or copper ions. In order to perform these functions, metals must be in oxidation states and coordination environments suitable for the reversible coordination of dioxygen. The iron porphyrin compound hemoglobin is found in red bloods of human beings and other animals.

Hemoglobin has the structure of heme iron with four iron porphyrin units combined with a globin protein. Dioxygen is transported in blood by being coordinated to ferrous ions in the hem iron unit. The Fe (II) ion is penta-coordinate with four nitrogen atoms of porphyrin and a nitrogen atom of the polypeptide histidine, and becomes hexa-coordinate when a dioxygen coordinates to it. The spin state of Fe (II) changes from high spin to low spin upon the coordination of dioxygen. The high spin Fe(II) is above the plane of porphyrin because it is too large to fit in the available space. When the Fe(II) ion becomes low spin upon dioxygen coordination, the size of the iron ion decreases and it just fits into the hole of the porphyrin molecule.

This molecular-level movement has attracted interest as an **allosteric effect** because it affects the whole protein through the histidine coordinate bond and governs the specific bond of a dioxygen molecule. Oxidation of the Fe(II) ion of a hem molecule is prevented by a macromolecular protein, and if the hem iron is taken out of the protein, Fe(II) ion is oxidized to Fe(III), and two porphyrin rings are bridged by a peroxide $\propto$-$O_2{}^{2-}$, which finally changes to a bridging $\propto$-O_2-structure.

When the hem is in this state, it loses the ability to coordinate to the dioxygen molecule. Based on this phenomenon, a synthetic porphyrin that is able reversibly to coordinate to a dioxygen by suppressing dimerization of the iron porphyrin has been developed, and was named the **picket fence porphyrin** after its three dimensional form.

(b) Nitrogen Fixation

The reaction which converts the nitrogen in air into ammonia is basic to all life. **Nitrogen fixation**, the reaction to fix atmospheric nitrogen to form ammonia, is carried out by *Rhizobium* in the roots of legumes or by bacteria in algae in an anaerobic atmosphere. All animals and plants, including mankind, were depended on biological nitrogen fixation as a source of nitrogen for protein and other compounds containing nitrogen before the invention of the Harber-Bosch process.

$$N_2 + 8\,H^+ + 8\,e^- + 16\,MgATP \longrightarrow 2\,NH_3 + H_2 + 16\,MgADP + 16\,Pi$$

(where Pi is an inorganic phosphate)

An enzyme named **nitrogenase** catalyzes this reaction. Nitrogenase contains iron-sulfur and iron-molybdenum sulfur proteins, and reduces dinitrogen by coordination and cooperative proton and electron transfers, while using MgATP as an energy source. Because of the importance of this reaction, attempts to clarify the structure of nitrogenase and to develop artificial catalysts for nitrogen fixation have continued for many years. Recently, the structure of an active center in nitrogenase called iron-molybdenum cofactor was clarified by single crystal X-ray analysis (Figure 4.26). According to this analysis, its structure has Fe_3MoS_4 and Fe_4S_4 clusters connected through S.

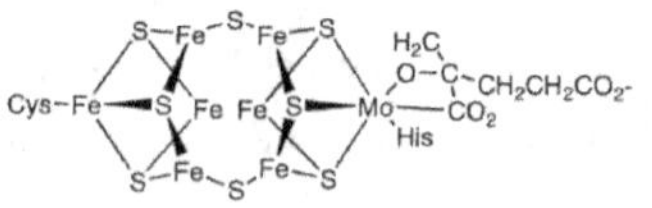

Figure 4.26: Structure of Fe-Mo Cofactor in Nitrogenase

It is believed that dinitrogen is activated by coordination between the two clusters. On the other hand, the portion called P cluster consists of two Fe_4S_4 clusters. The roles and reaction mechanism of both parts are not yet clear.

(c) Photosynthesis

The formation of glucose and dioxygen by the reaction of carbon dioxide and water is a skillful reaction using photoenergy and in which chlorophyll (Figure 4.27), which is a magnesium porphyrin and a manganese cluster complex, plays the central role. A chloroplast contains photosystem I (PSI) and photosystem II (PSII), which use light energy to reduce carbon dioxide and to oxidize water.

Chlorophyll is a fundamental component of PSI. Chlorophyll is a porphyrin complex of magnesium and is responsible for the green colors of leaves. It plays an important role in receiving light energy and transferring it to redox reaction systems. Chlorophyll is excited from the singlet ground state to the singlet excited state by light, the energy of the excited state is transferred to an acceptor within 10 ps, and the resultant energy reduces an iron-sulfur complex and is finally used for reduction of carbon dioxide in subsequent dark reactions. Since charge separation by photochemical excitation is the most important first stage, studies on photo induced electron transfer are have been actively performed using various kinds of porphyrin compounds as models of chlorophylls. PSI, which obtains oxidizing energy by electron transfer, converts ADP to ATP.

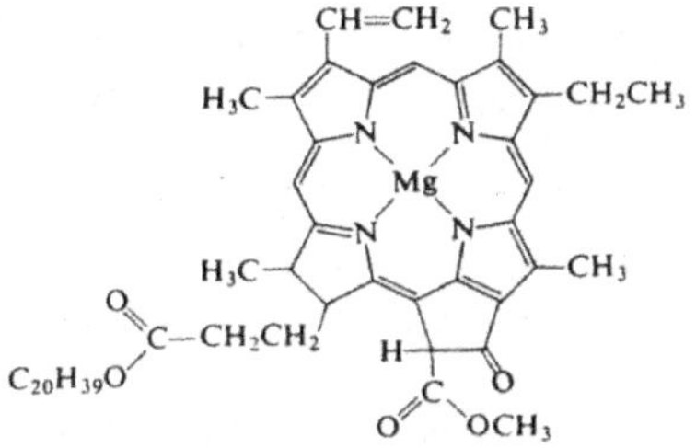

Figure 4.27: Chlorophyll *a*

On the other hand, the oxidized form of PSII oxidizes water through a chain of redox reactions of oxo cluster complexes of manganese, and generates oxygen. Since four electrons shift in the reduction of Mn(IV) to Mn (II) in this reaction, at least two manganese species are involved. Probably, a cluster complex which contains two Mn(II) and two Mn(IV) species mediates the electron transfer via four step reactions. However, the details of this reaction are as yet unclear because it is very difficult to isolate this cluster and to analyze its structure. The electron transfer stage is being studied at present by using various manganese complexes as model systems.

Photosynthesis is a very interesting research theme in bioinorganic chemistry as it involves a few metal ions, a porphyrin, sulfide and oxide clusters that constitute a cycle of subtle electron transfer and redox reactions, and generate oxygen gas by photolysis of water and produce carbohydrates from carbon dioxide by reductive dark reactions,. Recently, the reaction center of a photosynthetic bacteria was crystallized and J. Deisenhofer and his colleagues won a Nobel prize for its structural analysis (1988).

Exercise 1: Give examples of small molecules that are fundamentally important for living things.

"Answer" H_2O, O_2, N_2, CO_2.

Physical Properties

It is barely imaginable that materials based on the physical properties of solid inorganic compounds have played such decisive roles in present-day technology and industry. One may think that this field belongs more to material physics. However, apart from the theories of physical properties, the contribution of chemistry and chemists to the preparation of materials and their structural analysis has been greater than that of other branches of science. Material science is the application of the fundamental physical properties of materials such that basic theories and their applications converge. Therefore, by surveying the applications in such fields, the outlines of research themes and their purposes can be understood.

Important inorganic materials are surveyed from the chemical point of view by focusing on the relationship between preparation and isolation, and structure and physical properties.

(a) Electric Properties

A semiconductor is an electrical conductor with electrical resistance in the range of about 10^4 to 10^8 ohms. A typical semiconductor is a super-high grade silicon that is manufactured on a large scale and is widely used for information processing devices such as computers and energy conversion devices such as solar cells. VLSI (very large-scale integrated circuits) are printed on wafers made from almost defectless silicon single crystals with diameters of no less than 20 cm, prepared from polycrystalline silicon by the Czochralski method. Memory chips with a very high degree of integration as well as highly efficient computer chips have recently been realized.

In a short periodic table, silicon is a group IV element and has four valence-electrons. Although silicon semiconductors currently represent 90% or more of all semiconductors, isoelectronic 1:1 compounds of II-VI or III-V groups form compound semiconductors and are also used for optical or ultra high-speed electronic devices. For example, ZnS, CdS, GaAs, InP, *etc.* are typical compound semiconductors and the development of technologies to grow single crystals of these materials is remarkable. Light emitting diodes (LED) or semiconductor lasers are important applications of compound semiconductors.

As thin films of compound semiconductors are made by MBE (molecular beam epitaxy) or MOCVD (metallorganic chemical vapor deposition), special organometallic compounds, such as trimethyl gallium $Ga(CH_3)_3$ and trimethylarsenic $As(CH_3)_3$, which previously found little application, are now used industrially.

Superconductivity is a phenomenon of zero electrical resistance below a critical temperature, T_c, and was discovered in 1911 by Kamerlingh Onnes (1913 Nobel Prize for physics), who succeeded in liquefying helium during his experiments to measure the electrical resistance of mercury at ultra low temperatures. About 1/4 of the elements, such as Nb (T_c = 9.25 K), In, Sn, and Pb behave as superconductors and more than 1000 alloys and intermetallic compounds are also superconductors, but only Nb-Ti alloy (T_c = 9.5 K) and Nb_3Sn (T_c = 18 K) find application. Nb_3Sn, Nb_3Ge, V_3Ga, *etc.*, are cubic A-15 type compounds, in which transition metal atoms are aligned in chains, and interatomic distances are shorter than those in the crystalline bulk metal, raising the density of states of the conduction band and the critical temperature, T_c, of the compound.

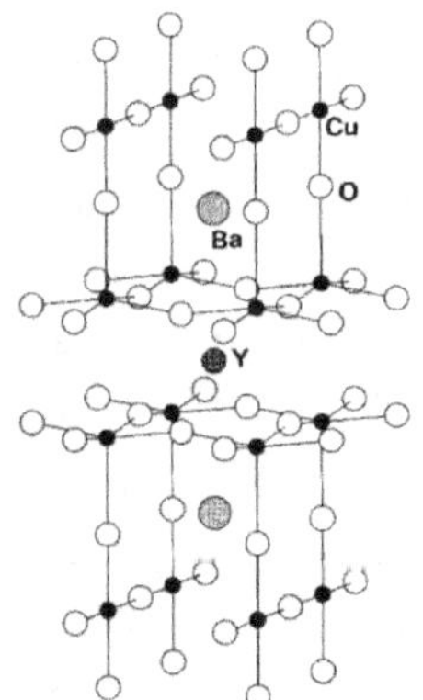

Figure 4.28: Structure of $YBa_2Cu_3O_{7-x}$

Among inorganic compound superconductors, chalcogenide compounds $M_xMo_6X_8$ (X = S, Se, Te, and M = Pb, Sn, *etc.*) of molybdenum called **Chevrel phases** and high-temperature superconductors of copper oxide derivatives, which J. G. Bednortz and K. A. Müller discovered in 1986 (1987 Nobel Prize for physics), have attracted attention. Chevrel phases have structures (refer to Section 4.4) in which hexanuclear cluster units of molybdenum are joined and the highest T_c is only 15 K of $PbMo_6S_8$, but the superconductive state is not broken even in strong magnetic fields. In the copper oxide system, more than 100 similar compounds have been prepared since the first discovery and the highest T_c so far discovered is 134 K. A typical compound, $YBa_2Cu_3O_{7-x}$, has a structure (Figure 4.28) in which CuO_5 square pyramids and CuO_4 planes are connected by corner-sharing, Ba and Y are inserted between them, and the oxygen content is non-stoichiometric.

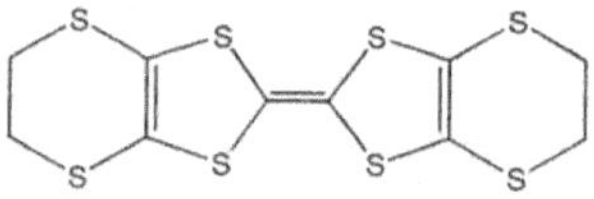

BEDT-TTF

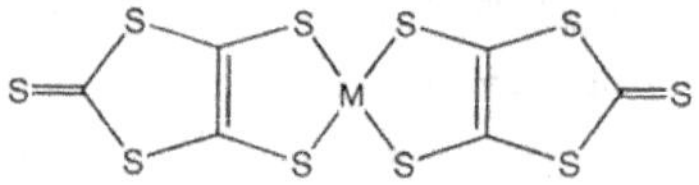

$[M(dmit)_2]^{2-}$ M = Ni, Pd

Figure 4.29: An Electron Donor and Acceptor in a Complex Superconductor

On the other hand, molecular superconductors have also been studied. Representative donor-acceptor complexes are composed of TTF and BEDT-TTF (Figure 4.29) as electron donors, and ClO_4^- or $[Ni(dmit)_2]^{2-}$ as electron acceptors. The first example of this kind of superconductor was discovered in 1980, and of the about 50 complexes known at present, the highest T_c is 13K. Recently (1991) fullerene C_{60} doped with alkali metals showed a T_c of about 30K.

Although thousands of superconductors are known, only a few of them find application. Because compound superconductors are very brittle; either it is difficult to make them into wires or only small single crystals are obtained. It will take considerable time before some of them find practical use. Therefore, mainly Nb-Ti wires are used as the superconducting magnets of analytical NMR, medical MRI (magnetic resonance imaging instrument) or maglev trains, *etc.* Efforts are concentrated on discovering materials that have suitable mechanical and other properties by cooperation between inorganic chemistry and solid-state physics.

Various metal oxides are used as thermistors (temperature sensitive resistance device), varistors (nonlinear resistance device), capacitors, *etc.* For example, $BaTiO_3$, with a perovskite structure, and $SrTiO_3$, *etc.* can be used for any of the above-mentioned purposes. Ionic conduction materials are also called solid electrolytes and α-AgI, β-Al_2O_3, stabilized zirconia (a part of Zr in ZrO_2 is replaced by Ca or Y), *etc.* are used in solid state batteries or fuel cells.

(b) Magnetism

Magnetic materials are divided into hard (permanent magnets) and soft magnetic materials. Permanent magnets are indispensable to machines using motors and MRI, which requires a high magnetic field. Japan has a strong tradition in the development of magnets, and has made many epoch-making magnetic materials for practical use. Alnico magnets with Fe, Ni, and Al as their main constituents, ferrite magnets composed of solid solutions of $CoFe_2O_4$ and Fe_3O_4, cobalt-rare earth magnets such as $SmCo_5$, and Nb-Fe-B magnets were especially significant achievements. Since soft magnetic materials are strongly magnetized in weak magnetic fields, they are most suitable for use as core materials in transformers. Hard magnetic properties are necessary for the stable maintenance of information whereas soft magnetic properties are required for recording and over-writing information in magnetic recording materials such as magnetic tapes, floppy disks, and hard disks. Although γ-Fe_2O_3 is a typical magnetic powder used for these purposes, Co^+ or crystalline CrO_2 is added to it to improve its magnetic properties. Recording materials as well as semiconductor devices are indispensable to our modern information-oriented society, and the role played by inorganic chemistry in the improvement of the performance of these materials is significant. Recently, ferromagnetism of organic compounds or metal complexes has been discovered, in which unpaired spins are aligned parallel in a molecule and coupled ferromagnetically. The study of molecular magnets has the subject of intensive investigation. Molecular design to couple paramagnetic metal complexes and to make spins parallel is an interesting subject in coordination chemistry.

(c) Optical Properties

Mainly inorganic substances are used as materials for optical applications. The optical fiber in particular has been used for optical communications on a large scale, and has had a major social influence in information communication. A necessary property of good optical glass materials is the transmission of information to distant places with little optical loss. Silica fibers are manufactured by lengthening silica glass rods produced from silica grains. The silica is made from ultra pure $SiCl_4$, which is oxidized in the vapor phase by an oxyhydrogen flame. As the optical loss along fibers obtained by this method has already reached its theoretical limit, fluoride glasses are being used in the search for materials with lower levels of loss.

Compound semiconductors such as GaP are widely used as laser light emitting diodes for optical communications, CD players, laser printers, *etc.* A high output YAG laser is made from neodymium-doped yttrium aluminum garnet, $Y_3Al_5O_{12}$, which is a double oxide of Y_2O_3 and Al_2O_3. Single crystals, such as lithium niobate, $LiNbO_3$, are used for changing wavelength of light by means of SHG (second harmonic generation) of nonlinear optics phenomena.

Structure-function Correlation

Since all the naturally occurring elements have been discovered, various bonding modes are established and the structures of compounds can be readily determined, studies of the chemical properties of inorganic compounds will give way to studies of reactions and physical properties. The synthesis of new compounds and the elucidation of structure-function correlations will be the foundations of these studies, although the end is distant.

It is considerably difficult quantitatively to explain the thermal stability of a known inorganic compound using our present knowledge of theoretical chemistry and it is almost impossible fully to design compounds by a rational method. Although the selectivity of a catalytic reaction can be explained to some extent, the theoretical calculation of a reaction rate remains difficult. The relation between superconductivity and structure is not understood well, and critical temperatures cannot be predicted. Many of the structures and functions of the metalloenzymes that are the basis of biological activities are unknown. The research problems confronting the next generation of inorganic chemists are extensive, and novel solutions can be anticipated.

Questions

Part A

1. Write a balanced equation for the preparation of diborane.
2. Write a balanced equation for the preparation of triethylphosphine.
3. Write a balanced equation for the preparation of osmium tetroxide.
4. Describe the basic reaction of the phosphomolybdate method used for the detection of phosphate ions.

5. Draw the structure of anhydrous palladium dichloride and describe its reaction when dissolved in hydrochloric acid.

6. Describe the reaction of anhydrous cobalt dichloride when it is dissolved in water.

7. Draw the structure of phosphorus pentafluoride.

8. Write a catalytic reaction cycle of the hydroformylation reaction which uses $[RhH(CO)(PPh_3)_3]$ as a catalyst.

9. Describe differences between ammonia synthesis by the Harber-Bosch process and biological nitrogen fixation reactions.

10. A-15 type intermetallic compounds such as Nb_3Sn are cubic crystals with the A_3B composition . Consider how to locate each atom in such a unit cell.

Part B

1. Explain in detail about non metals

2. Explain in detail about reaction and physical properties of inorganic elements

3. Brief : Structure-function correlation

CHAPTER 5

APPLICATIONS OF CHEMISTRY

5.1. Thermo Chemistry

The Concepts of Heat and Energy

Energy is defined as the ability to do work and supply heat.

In general the discussion of energy is divided into the concepts of kinetic energy and potential energy.

Kinetic Energy

Kinetic energy is the energy an object has due to its motion and is defined by the following formula.

$$KE = \tfrac{1}{2}\, mv^2$$

Where

KE = kinetic energy

m = mass in kg

v = velocity in m/s

The unit of kinetic energy is the **Joule** which is defined as the amount of energy a 2kg object has if it is moving at 1 m/s.

The units for a Joule come directly from the formula above.

$$KE = \tfrac{1}{2}\, (2kg)(1m/s)^2 = kg\ m^2\ s^{-2}$$

Potential Energy

Potential energy is the energy stored in a stationary object due to its position or condition.

The units of potential energy are the same as those for kinetic energy. If you drop a 2kg object and determine that it achieved 2 Joules of kinetic energy, then it is defined that the object had contained 2 Joules of potential energy. In the case of potential energy mentioned above it is the Earth's gravity that gives the object its potential energy. Other forces such as electrical, mechanical, and magnetic forces can also give potential energy to objects.

Some Examples of Kinetic and Potential Energy

A boulder at the top of a mountain has *potential* energy. When it begins to roll down the mountain its energy becomes *kinetic*.

The gasoline in your car has *potential* energy that is turned into *kinetic* energy when it is combusted. If you stretch a rubber band you add *potential* energy to it that becomes *kinetic* energy when it is released. A snowflake has *kinetic* energy which is converted to *potential* energy when it settles to the ground.

Calculating Heat Content
Kinetic Energy of a System of Components

Rather than looking at a single object in motion let's look at a system of components, such as atoms or molecules, that are all in motion and colliding with one another.

When we wish to express the kinetic energy realized by these motions and collisions, we refer to it as *heat*.

The unit for heat is the *Joule* just as it is for work. However, a more common unit, the *calorie*, is also used.

The calorie is defined as the amount of energy required to raise one gram of water 1°C and has the symbol "*cal*". A dietary calorie is equal to 1,000 calories and has the symbol "*Cal*".

$$1 \text{ calorie} = 4.184 \text{ Joules}$$

Exercise 1

A 30 g portion of fat contains 1100kJ of energy. How many dietary calories would that be?

$$\frac{1100 \text{ kJ}}{} = \qquad \text{Cal}$$

$$\frac{1100 \text{ kJ}}{} \left| \frac{10^3 \text{J}}{1 \text{ kJ}} \right. = \qquad \text{Cal}$$

Convert kJ to J.

$$\frac{1100 \text{ kJ}}{} \left| \frac{10^3 \text{J}}{1 \text{ kJ}} \right| \frac{1 \text{ cal}}{4.181 \text{ J}} = \qquad \text{Cal}$$

Convert J to cal.

$$\frac{1100 \text{ kJ}}{} \left| \frac{10^3 \text{J}}{1 \text{ kJ}} \right| \frac{1 \text{ cal}}{4.181 \text{ J}} \left| \frac{1 \text{ Cal}}{10^3 \text{ cal}} \right. = 263 \text{ Cal}$$

Convert cal to Cal.

Problem 5.1

A 25 g sample of soda pop contains 2500 kJ of energy. How many Calories does the soda pop contain?

$$\frac{2500 \text{ kJ}}{} \Big| = \qquad \text{Cal}$$

$$\frac{2500 \text{ kJ}}{} \Big| \frac{10^3 \text{ J}}{1 \text{ kJ}} = \qquad \text{Cal}$$

Convert kJ to J.

$$\frac{2500 \text{ kJ}}{} \Big| \frac{10^3 \text{ J}}{1 \text{ kJ}} \Big| \frac{1 \text{ cal}}{4.181 \text{ J}} = \qquad \text{Cal}$$

Convert from J to cal.

$$\frac{2500 \text{ kJ}}{} \Big| \frac{10^3 \text{ J}}{1 \text{ kJ}} \Big| \frac{1 \text{ cal}}{4.181 \text{ J}} \Big| \frac{1 \text{ Cal}}{10^3 \text{ cal}} = 598 \text{ Cal}$$

Convert from cal to Cal.

Problem 5.2

How many kilojoules are in a candy bar containing 750 Calories?

$$\frac{750 \text{ Cal}}{} \Big| = \qquad \text{kJ}$$

$$\frac{750 \text{ Cal}}{} \Big| \frac{10^3 \text{ cal}}{1 \text{ Cal}} = \qquad \text{kJ}$$

Convert Cal to cal.

$$\frac{750 \text{ Cal}}{} \Big| \frac{10^3 \text{ cal}}{1 \text{ Cal}} \Big| \frac{4.184 \text{ J}}{1 \text{ cal}} = \qquad \text{kJ}$$

Convert cal to J.

$$\frac{750 \text{ Cal}}{} \Big| \frac{10^3 \text{ cal}}{1 \text{ Cal}} \Big| \frac{4.184 \text{ J}}{1 \text{ cal}} \Big| \frac{1 \text{ kJ}}{10^3 \text{ J}} = 3180 \text{ kJ}$$

Convert j to kJ.

Problem 5.3

Calculate the number of kilojoules in a potato containing 105 kilocalories.

$$\frac{105 \text{ kcal}}{} = \quad \text{kJ}$$

$$\frac{105 \cancel{\text{ kcal}}}{} \left|\frac{10^3 \text{ cal}}{1 \cancel{\text{ kcal}}}\right. = \quad \text{kJ}$$

Convert kcal to cal.

$$\frac{105 \cancel{\text{ kcal}}}{} \left|\frac{10^3 \cancel{\text{ cal}}}{1 \cancel{\text{ kcal}}}\right| \frac{4.184 \text{ J}}{1 \cancel{\text{ cal}}} = \quad \text{kJ}$$

Convert cal to J.

$$\frac{105 \cancel{\text{ kcal}}}{} \left|\frac{10^3 \cancel{\text{ cal}}}{1 \cancel{\text{ kcal}}}\right| \frac{4.184 \cancel{\text{ J}}}{1 \cancel{\text{ cal}}} \left|\frac{1 \text{ kJ}}{10^3 \cancel{\text{ J}}}\right. = 439 \text{ kJ}$$

Convert from j to kJ.

Exothermic and Endothermic Reactions

When observing the heats of chemical reactions it is important to know which way the heat is flowing in the chemical reaction.

Is the heat being generated by the chemical reaction? Is heat required in order for the chemical reaction to proceed?

The following definitions are agreed upon conventions describing the flow of heat in chemical reactions.

Exothermic Reactions

Chemical reactions that ***release heat to the surroundings*** are considered to be exothermic reactions. Let's look at the familiar combustion reaction of propane (C_3H_8).

$$C_3H_8 + 5O_2 \rightarrow 3CO_2 + 4H_2O$$

In truth this reaction is missing one very important product.

$$C_3H_8(g) + 5O_2(g) \rightarrow 3CO_2(g) + 4H_2O(g) + \textbf{\textit{Heat}}$$

When one mole of propane is burned it releases 2028.8 kJ of heat to the surroundings.

Due to convention the heat of reaction for the combustion of propane is expressed as -2028.8 kJ mol^{-1}. The negative sign is from the perspective of the reactants. The propane and the oxygen **lost heat** as a consequence of the combustion reaction.

Endothermic Reactions

Chemical reactions that need to **absorb heat from the surroundings** in order to proceed are known as endothermic reactions.

Let's look at the formation of nitrogen monoxide (NO) form nitrogen (N_2) and oxygen (O_2),

$$N_2(g) + O_2(g) \rightarrow 2NO(g)$$

Again this chemical equation does not tell the entire story. The reaction will not proceed unless energy, in the form of heat, is added to the reaction.

$$N_2(g) + O_2(g) + \textbf{\textit{Heat}} \rightarrow 2NO(g)$$

It is necessary for N_2 and O_2 to absorb 90.37 kJ mol^{-1} in order to react to form one mole of NO. In the above balanced chemical equation twice that value, 180.74 kJ, were required to form two moles of NO.

By convention the heat of reaction for the formation of NO from N_2 and O_2 is 90.37 kJ mol^{-1}. This number is positive since the N_2 and O_2 are seen to have gained heat.

Enthalpy Calculations: Hess' Law

When an object or system of objects are heated more is going on than just the heating.

If you heat a beaker of water that is open to ambient pressure the volume of the water will increase. This is a form of work know as PΔv work.

Rather than deal with this component of thermochemistry directly, a new value known as **enthalpy (H)** was developed.

Standard enthalpies of formation for many substances were experimentally determined and listed in tables. It is easy to use these listed values to calculate the change in enthalpy for a reaction using the following formula.

Hess' Law

Hess' Law uses standardized enthalpies of formations for reactants and products in order to calculate the change in enthalpy for a chemical reaction.

$$6H_{rxn} = \Sigma H°\text{form-prod} - \Sigma H°\text{form-react}$$

Where "Δ" signifies a change in

"Σ" signifies a summation of

"$°$" signifies a standardized value

"rxn" signifies the chemical reaction

Hess' Law simply states that if you subtract the standard energies (enthalpies) of formation of the reactants from the standard energies of formation of the products that whatever is left is the energy change for that reaction.

If the value is negative, that means that energy is released to the surroundings and the reaction is **exothermic.**

If the value is positive, that means that the energy is absorbed from the environment and the reaction is **endothermic.**

The following table lists standard enthalpies of formations for some compounds. This table can be used to work the exercises that follow.

Table 5.1: Enthalpies of Formation for Selected Substances

Substance	Formula	ΔH_{rxn} (kJ mol^{-1})
Water – liquid	$H_2O(l)$	-284.83
Water – gas	$H_2O(g)$	-241.816
Carbon Dioxide	CO_2	-393.5
Carbon Monoxide	CO	-110.525
Ammonia	NH_3	-80.8
Ammonium Chloride	NH_4Cl	-314.55
Oxygen	O_2	0
Nitrogen	N_2	0
Hydrogen Chloride	HCl	-167.2
Methane	CH_4	-17.9
Ethane	C_2H_6	-20.0
Acetylene	C_2H_2	+54.2
Propane	C_3H_8	-25.0

States of Matter

The three principle states of matter are **solid, liquid,** and **gas.** Below are listed the basic definitions that differentiate these three states of matter from one another.

Solids

Solids are materials that have a definite shape and volume. The arrangement of atoms or molecules in a solid are fixed. The movement of the atoms or molecules in a solid are very slow.

Liquids

Liquids are materials that have a definite volume but an indefinite shape. The arrangement of liquid atoms or molecules is random. The shape of a liquid is determined by the container holding the liquid. The movement of atoms and molecules in a liquid is faster than they are in a solid.

Gases

Gases are materials that have an indefinite shape and volume. Both the shape and the volume of a gas are determined by the shape and the volume of the container holding the gas. The arrangement of atoms or molecules in a gas is random. The movement of atoms or molecules in a gas is very rapid.

Changes of States in Matter

Figure 5.1 indicates the six changes of state for matter

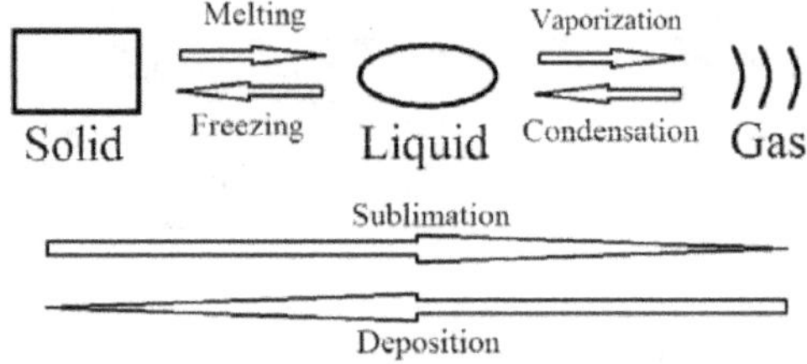

Figure 5.1: Changes of State for Matter

5.2.   Electrochemistry

The Galvanic Cell

If you immerse two dissimilar metals into an electrolytic or ionic solution, a reduction oxidation reaction will occur that will result in a transfer of electrons between the two metals. This results in a potential difference or voltage between the two metals. You will then have created a *galvanic cell* more commonly known as a battery.

The Daniell cell is a type of galvanic cell that utilizes the oxidation of zinc metal (Zn) to zinc ions (Zn^{2+}) and the reduction of copper ions (Cu^{2+}) to copper metal (Cu).

Overall reaction: $Zn(s) + Cu^{2+}(aq) \rightarrow Zn^{2+}(aq) + Cu(s)$

Oxidation half-reaction (anode): $Zn(s) \rightarrow Zn^{2+}(aq) + 2e^-$

Reduction half-reaction (cathode): $Cu^{2+}(aq) + 2e^- \rightarrow Cu(s)$ Remember that *oxidation* is a *loss of electrons* and *reduction* is *a gain of electrons*.

As seen in the above half-reactions, *oxidation* occurs at the *anode* and *reduction* occurs at the *cathode*.

A Daniell cell can easily be constructed. Into two beakers containing a solution of sodium sulfate (Na_2SO_4), insert copper and zinc strips. Prepare a salt bridge between the two beakers. Wire the two metal strips in series with a test lamp. The lamp will glow with the current flow from the potential difference between the two metals.

Figure 5.2 Shows a typical Daniell cell.

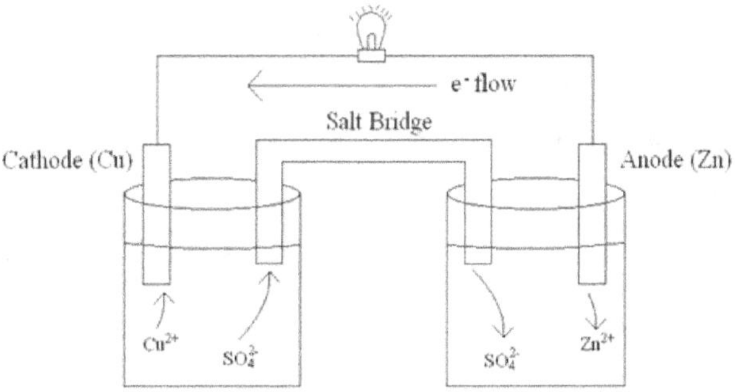

Figure 5.2: A Typical Daniell Cell

I'm sure everyone has seen the popular novelty toy where a fruit or vegetable is used as an electrolytic source to light a small light bulb.

Two wires of dissimilar metals, such as copper and zinc, are wired in series with the light bulb. The ends of the wires are inserted into the fruit or vegetable and a current sufficient to light the bulb is generated.

In the fruit or vegetable both the electrons and ionic species are free to flow in what is essentially a completed circuit.

The purpose of the salt bridge in the Daniell cell is to bridge the two separate containers to complete the electrical circuit.

Shorthand Notation for Daniell Cells

The following shorthand notation has been adopted for Daniell cells, once again using the copper-zinc cell for demonstration.

$$Zn(s) \,|\, Zn^{2+}(aq) \,\|\, Cu^{2+}(aq) \,|\, Cu(s)$$

$$e^{-} \text{ flow} \longrightarrow$$

The single vertical lines represent phase boundaries and the double vertical lines represent the salt bridge.

What's Going On in a Typical Battery?

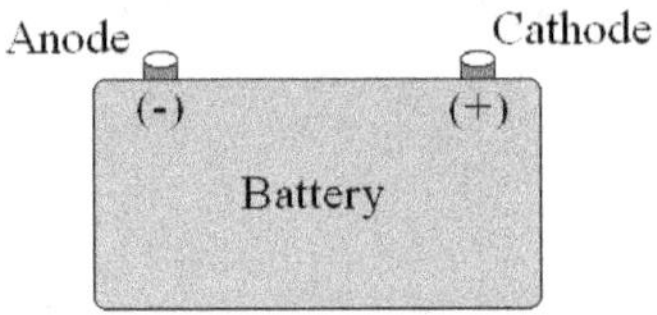

Figure 5.3: A Typical Daniell Cell

Let's break down the reduction oxidation chemistry between the electrodes.

The Anode

Oxidation occurs at the anode. Since oxidation is a loss of electrons it is at the anode that electrons are generated. Therefore, the anode has a negative sign.

The Cathode

Reduction occurs at the cathode. Since reduction is a gain of electrons it is to the cathode that electrons flow. Therefore, the cathode is positive.

Standard Reduction Potentials

It must first be noted that neither a reduction potential nor an oxidation potential can be measured on a single electrode. Potentials can only be measured **between two electrodes**.

Early chemists decided that these potentials would be measured against an arbitrary standard electrode and recorded as **reduction potentials.**

The Standard Hydrogen Electrode (S.H.E.)

The **standard hydrogen electrode** is defined as platinum (Pt) in contact with hydrogen (H_2) gas, at 1 atmosphere, being bubbled through an aqueous solution that is 1 molar in hydrogen ions (H^+) at 25°C.

In either direction the reaction is viewed the electric potential of the cell was arbitrarily standardized as exactly zero volts.

$$2H^+(aq, 1\ M) + 2e^- \rightarrow H_2 \rightarrow (g,\ 1\ atm) \qquad E° = 0\ V$$

$$H_2(g, 1\ atm) \rightarrow 2H^+(aq, 1\ M) + 2e^- \qquad E° = 0\ V$$

With this standardization accomplished the standard reduction potentials of half cell reactions could then be compared to one another.

Electric Cell Potentials

Now that we have established standardized half-cell potentials it should be relatively easy to calculate the potential of a complete cell.

The following equation establishes the potential for a Daniell cell.

$$E°_{cell} = E°_{oxidation} + E°_{reduction}$$

Doing cell potential calculations comes with one very important caveat. The half-cell potential of the species being oxidized in the cell's oxidation-reduction reaction is viewed as the *negative value* of that half-cell's listed standard reduction potential.

Example

Calculate the cell potential for the copper-zinc (Cu-Zn) cell that we have looked at in the beginning of this section.

The copper (Cu) is being reduced with a standard reduction potential of 0.34 V.

The zinc (Zn) is being oxidized with a standard reduction potential of -0.76 V.

$$E°_{cell} = E°_{oxidation} + E°_{reduction}$$

$$E°_{cell} = -(-0.76\ V) + 0.34\ V = 1.10\ V$$

Problem 5.4

Calculate the cell potential of a copper-cadmium (Cu-Cd) cell.

The copper (Cu) is being reduced with a standard reducing potential of 0.34 V.

The cadmium (Cd) is being oxidized with a standard reduction potential of -0.40 V.

$$E°cell = -(-0.40) + 0.34\ V = 0.74\ V$$

Problem 5.5

Calculate the cell potential of a chlorine-iron (Cl-Fe) cell.

The chlorine (Cl) is being reduced with a standard reducing potential of 1.36 V.

The iron (Fe) is being oxidized with a standard reduction potential of -0.45 V.

$$E°cell = -(-0.45) + 1.36 \text{ V} = 1.81 \text{ V}$$

Problem 5.6

Calculate the cell potential of a silver-aluminum (Ag-Al) cell.

The silver (Ag) is being reduced with a standard reduction potential of 0.80 V.

The aluminum (Al) is being oxidized with a standard reduction potential of -1.66 V.

$$E°cell = -(-1.66) + 0.80 \text{ V} = 2.46 \text{ V}$$

Problem 5.7

Calculate the cell potential of a hydrogen ion-lithium (H^+-Li) cell.

The hydrogen ion (H^+) is being reduced with a standard reducing potential of 0 V. The lithium (Li) is being oxidized with a standard reduction potential of -3.04 V.

$$E°cell = -(-3.04) + 0 \text{ V} = 3.04 \text{ V}$$

Electrolysis

Up to this point, in this chapter, we have discussed only the galvanic cell. In a galvanic cell spontaneous chemical reactions occur which, through oxidation-reduction, result in a potential.

But what if you wanted to accomplish a nonspontaneous reaction for a specific purpose? You could power a cell with an external source of electricity to force the nonspontaneous reaction to occur. This would then be an ***electrolytic cell*** accomplishing ***electrolysis***.

This happens every time you start your car. With the car not running the battery provides electrical power to the starter to start the engine. This is done through the use of a galvanic cell. Once the engine is running the alternator uses the battery as an electrolytic cell to run the galvanic reaction backwards and charge the battery.

Electrolysis of Water

Electrolysis of water is used to produce hydrogen gas (H_2) and oxygen gas (O_2). The following chemical equations describe the electrolysis of water:

Oxidation occurs at the anode: $2H_2O(l) \rightarrow O_2(g) + 4H^+(aq) + 4e^-$

Reduction occurs at the cathode: $4H_2O(l) + 4e^- \rightarrow 2H_2(g) + 4OH^-(aq)$

Electroplating

Electroplating is an electrolysis based technique that deposits one metal on the surface of another. Silver can be electroplated onto steel to form a silver-plated steel object.

Corrosion

Corrosion, by definition, is the oxidative decomposition of a metal. But as we all should know, oxidation cannot exist without reduction. Let's look at the most well known occurrence of corrosion the rusting of iron. The corrosion of iron will only occur in the presence of oxygen (O_2) and water (H_2O).

When water is in contact with iron (Fe), in the presence of oxygen (O_2), a galvanic cell is formed where oxidation occurs at an anodic portion of the surface that is shielded from oxygen (O_2) and oxygen (O_2) is reduced at a nearby cathodic region of the surface.

The following oxidation-reduction equations apply:

Anode:	$Fe(s) \rightarrow Fe^{2+}(aq) + 2e^-$	$E° = -0.45$ V
Cathode:	$O_2(g) + 4H^+(aq) + 4e^- \rightarrow 2H_2O(l)$	$E° = 1.23$ V

Protection from Corrosion

To protect iron surfaces from corrosion it is common to connect a **sacrificial anode** to the iron (Fe).

The choice of the material to protect iron (Fe) simply needs to be a metal that is more readily oxidized than iron (Fe).

Ships at sea will attach large blocks of zinc (Zn) to the iron (Fe) of the ship to act as sacrificial anodes to protect the iron (Fe).

Steel is often dipped in molten zinc (Zn) to prevent corrosion in a process known as **galvanization**.

Batteries

This section will present the oxidation-reduction chemistry of batteries that are common in everyday use.

Dry Cell Batteries

A conventional dry cell battery consists of a zinc (Zn) can acting as an anode and a graphite rod acting as a cathode. The graphite rod is coated with a manganese oxide (MnO_2) and carbon black paste.

The electrolyte is a damp paste consisting of ammonium chloride (NH_4Cl) and zinc chloride ($ZnCl_2$).

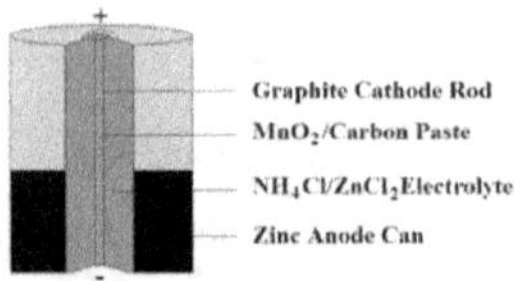

Figure 5.4: A Dry Cell Battery

Lead Storage Batteries

Your car battery is a lead storage battery. In fact it is actually six galvanic cells, producing 1.924 V each, connected in series to produce nearly twelve volts.

Lead ($Pb(s)$) is oxidized at the anode into lead sulfate $PbSO_4(s)$. Lead oxide ($PbO_2(s)$) is reduced at the cathode to lead sulfate $PbSO_4(s)$.

Anode

$$Pb(s) + HSO_4^-(aq) \rightarrow PbSO_4(s) + H^+(aq) + 2e^- \quad E° = -(-0.296)\ V$$

Cathode

$$PbO2(s) + 3H^+(aq) + 2e- \rightarrow PbSO_4(s) + 2H2O(l) \quad E° = 1.628V$$

The total cell voltage is 1.924 V. The total voltage of the battery is 11.544 V.

Nickel-Cadmium Batteries

Nickel-cadmium (or ni-cad) batteries are of popular use as dry cell batteries because they are rechargeable. Cadmium ($Cd(s)$) is oxidized at the anode. Nickeloxyhydroxide ($NiO(OH)$) is reduced at the cathode.

Anode

$$Cd(s) + 2OH^-(aq) \rightarrow +2e^- \quad\quad E° = -(-0.40)\ V$$

Cathode

$$NiO(OH)(s) + H_2O(l) + e^- \rightarrow Ni(OH)_2(s) + OH^-(aq) \quad E°=0.083\ V$$

The total voltage for a nickel-cadmium battery is 1.23 V.

Nickel-Hydride Batteries

Cadmium is very expensive and a toxic heavy metal. Nickel-hydride batteries present a more economical and environmentally safe alternative to cadmium batteries. A metal hydride ($MH(s)$) is oxidized at the anode. Nickeloxyhydroxide is reduced at the cathode.

Anode

$$MH(s) + OH^- \rightarrow M(s) + H_2O(l) + e^-$$

Cathode

$$NiO(OH)(s) + H_2O(l) + e^- \rightarrow Ni(OH)_2(s) + OH^-(aq)$$

The nickel-hydride battery produces approximately the same 1.2 volts as the nickel-cadmium battery.

Lithium-Ion Batteries

Since lithium is both a light weight metal and a very strong reducing agent, lithium-ion batteries are light weight and can produce just over 3 volts per cell.

The lithium-ion battery contains no lithium metal ($Li(s)$) as such. The anode is graphite that has been impregnated with lithium ions, $Li_xC_6(s)$. The cathode is cobalt oxide also containing lithium ions, $Li_{1-x}CoO_2(s)$.

Anode

$$Li_xC_6(s) \rightarrow xLi^+(aq) + 6C(s) + xe^-$$

Cathode

$$Li_{1-x}CoO_2(s) + xLi^+(aq) + xe^- \rightarrow LiCoO_2(s)$$

Fuel Cells

A fuel cell, like other batteries, is a galvanic cell with one distinct difference. The cell does not contain the reactants for the oxidation-reduction reaction.

The reactants of a fuel cell are fed to the cell on a continuous basis while the products of this reaction are channeled out of the cell.

The chemical process of a hydrogen-oxygen fuel cell, as shown in Figure is essentially the reaction of converting hydrogen (H_2) and oxygen (O_2) to water (H_2O).

Hydrogen is oxidized at the anode. Oxygen is reduced at the cathode. A hot solution of potassium hydroxide (KOH) acts as the electrolyte.

Anode

$$2H_2(g) + 4OH^-(aq) \rightarrow 4H_2O(l) + 4e^- \qquad E° = 0\,V$$

Cathode

$$O_2(g) + 2H_2O(l) + 4e^- \rightarrow 4OH^-(aq) \qquad E° = 1.18V$$

Of course the standard reduction potential of hydrogen (H_2) held in reference to the standard hydrogen electrode is 0 V. The total potential is the standard reduction potential of oxygen (O_2) of 1.18 V. It should be noted that the hydrogen-oxygen fuel cell is considerably eco-friendly. Consider that the "exhausts" of the cell are water and excess oxygen.

Figure 5.5 shows a diagram of a hydrogen-oxygen fuel cell.

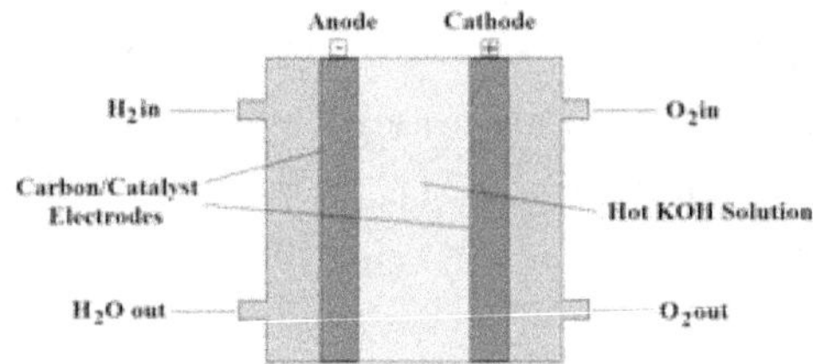

Figure 5.5: A Hydrogen-Oxygen Fuel Cell

5.3. Nuclear Chemistry

Nuclear Reactions

A nuclear reaction is a reaction where a change in the identity of the nucleus has occurred.

Look at the following reaction of carbon-14 in the atmosphere when exposed to cosmic radiation.

$$^{14}_{6}C \longrightarrow {}^{14}_{7}N + {}^{0}_{-1}e + \overline{\nu}_e$$

Here one of the neutrons of carbon-14 has been changed into a proton through the ejection of an electron and an antineutrino.

Distinctions of Nuclear Reactions

1. The nucleus is altered.

2. Different isotopes behave differently.

3. Reaction rate is unaffected by temperature, pressure, or the addition of a catalyst.

4. The reaction is the same whether the atom is uncombined or combined in a   molecule.

5. The energy change is much greater than that of a chemical reaction.

Types of Radiation

Alpha (α) Radiation

The first of the four major types of radiation to be described is *alpha (α)* radiation.

The alpha *particle* is essentially the nucleus of the helium atom. A decay of uranium-238 may be described by the following nuclear reaction equation.

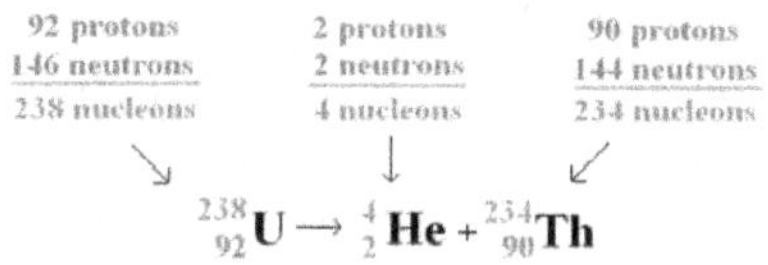

$$^{238}_{92}U \longrightarrow {}^{4}_{2}He + {}^{234}_{90}Th$$

Beta (β) Radiation

The second type of radiation to be described is *beta (β)* radiation. Beta particles are essentially electrons and have the symbol indicating a charge but no (appreciable) mass.

During the radioactive decay of iodine (I) to xenon(Xe) a neutron decays to a proton and emits an electron.

$$^{131}_{53}I \longrightarrow {}^{131}_{54}Xe + {}^{0}_{-1}e$$

Gamma (γ) Radiation

The third type of radiation is *gamma (γ)* radiation. Gamma radiation is simply very high electromagnetic radiation. It is unaffected by magnetic fields and has no mass.

Gamma radiation often accompanies the emission of β and α radiation as a release of energy. Gamma radiation is nearly always omitted when writing nuclear equations because it changes neither the mass number nor the atomic number of the product nuclei.

Positron (β⁺) Radiation

Positron (β⁺) Radiation

A fourth type of radiation is **positron (β+)** emission. A positron is essentially a **positively charged electron**.

During the natural decay of potassium (K) to argon (Ar) a proton is converted to a neutron and ejects a positron.

$$\begin{array}{ccc} \text{19 protons} & \text{18 protons} & \\ \underline{\text{21 neutrons}} & \underline{\text{22 neutrons}} & \textbf{+1 charge} \\ \text{40 nucleons} & \text{40 nucleons} & \text{0 nucleons} \end{array}$$

$$^{40}_{19}\text{K} \rightarrow {}^{40}_{18}\text{Ar} + {}^{0}_{1}\text{e}$$

Measurement of Radiation and Radioactive Dose

A **radiation measurement** is a measure of the rate of radiation emission from a radioactive source. A **radiation dose measurement** is a measure of biological effect on soft human or mammal tissue.

Radiation Measurement Units

Radiation is measured in Becquerel (Bq) and Curie (Ci).

A Becquerel (Bq) is defined as one event of radiation emission per second. The Curie (Ci), a vastly larger number, is equal to 37,000 megaBq.

Radiation Dose Measurement Units

The SI unit for radiation dose is the **Gray (Gy)** and is equal to the absorbed dose of 1 Joule per kilogram.

Another common unit of radiation dose is the **rad** and is equal to the absorbed dose of 0.01 Joule per kilogram or 0.01 Gy.

A **dose equivalent** is a special measurement of radiation dose that has been multiplied by special quality factors to account for different types of radiation and different parts of the body.

Radiation Dose Equivalent Measurement Units

The **Sievert (Sv)** is the SI unit of dose equivalence and is equal to the absorbed dose in Gy multiplied by special factors.

Another unit of dose equivalence is the **rem** and is equal to the absorbed dose in rad multiplied by special factors. One rem is equal to 0.01 Sv.

Radioactive Decay

Radioactive Decay Rates

Just like all other processes in chemistry if the rate can be characterized or calculated it will be.

Radioactive decay is a **first order process** whose rate is proportional to the number of radioactive nuclei (**N**) in a sample.

The decay rate is characterized by the following.

$$\text{Decay Rate} = k \times N$$

Where **k** is a rate constant called the **decay constant**.

Half-Life

Like all first-order processes, radioactive decay is characterized by a half-life ($t_{1/2}$), the time required for the number of radioactive nuclei to drop to half of the original value.

A general formula for half-life is calculated by assuming that N = ½ N0. ln(½N0/N0) = -k/t1/2 then ln½ = -ln2 = -kt1/2 therefore

$$t_{1/2} = ln2/k.$$

Note that the large number for carbon-14 is why it is so good at dating carbonaceous materials. Also, the immensely large number for uranium-238 is why the nuclear waste problem is so immense.

Fission

Mass Defects and Binding Energy

Binding energy (ΔE) is the energy that holds a nucleus together. If we force that nucleus apart (**fission**) there is a noticeable mass defect (Δm) that relates to the energy released according to the following formula.

$$dE = dmc^2$$

This now famous formula was what drove the scientists in the Manhattan Project forward to obtain the massive amounts of energy when the "missing" mass in the fission reaction was converted to energy.

Example

Let's look at a typical fission of uranium-235.

Subtract the accurate masses of the products from the reactants.

U-235 = 235.0439 amu   Kr-91 = 90.9234 amu

Ba-142 = 141.9164 amu   n = 1.00865 amu

(235.0439 + 1.00865) – (90.9234 + 141.9164 + 3.02595) = 0.1868 amu

Atomic masses directly predict molar masses. Therefore, the molar mass is 0.1868 g/mole.

$$\Delta E = \Delta mc^2 = (0.1868 \text{ g/mol})(10^{-3} \text{ kg/g})(3.00 \times 10^8 \text{ m/s})^2$$

$$= 1.68 \times 10^{13} \text{ kg} \bullet \text{m}^2/(\text{s}^2 \bullet \text{mol}) = \textbf{1.68} \times \textbf{10}^{\textbf{13}} \textbf{\textit{ J/mol}}$$

This massive amount of energy per mole of material absolutely dwarfs the amount of energy per mole you could get out of combusting 0.1868 grams of combustible material.

In Figure 5.6 a fast neutron fissions with an atom of U-238 which gives birth to, among other radioactive isotopes, approximately three fast neutrons per fission.

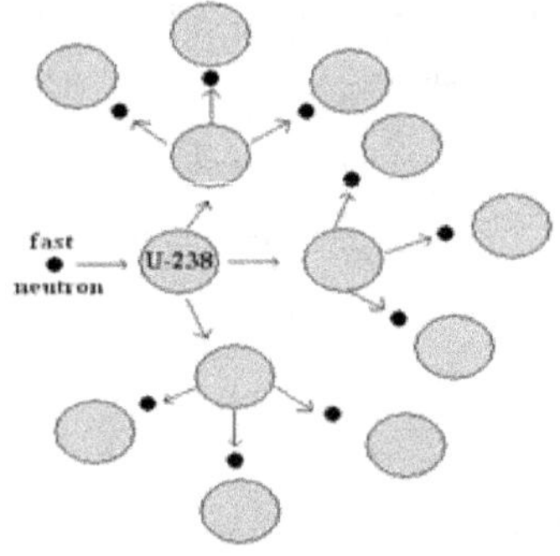

Figure 5.6: Fission of U-238

Since U-238 needs fast neutrons to fission and gives birth to fast neutrons, the uncontrollable nuclear fission reaction can take place.

U-235, on the other hand, requires a thermal (or slowed) neutron in order to fission yet it also gives birth to approximately three fast neutrons per fission.

The way the fast neutrons of U-235 are slowed down is by letting them collide with things that are nearly identical in size in a process know as inelastic scattering.

The particles used to effect the inelastic scattering of the fast neutrons are the hydrogen (H) atoms in a water (H_2O) molecule. With each collision between a fast neutron and a hydrogen (H) atom the momentum of the fast neutron is decreased. When the fast neutron has lost sufficient momentum it can then fission with a U-235 atom. For this reason the water is considered to be a ***moderator***.

Fusion

Nuclear Fusion

Just as heavy atoms, such as uranium, release vast amounts of energy when they undergo fission, very light atoms, such as the isotopes of hydrogen, also release enormous amounts of energy when they are forced together (***fusion***). It is believed that helium is formed in the sun through the following reactions.

$$_1^1H + {}_1^1H \rightarrow {}_1^2H + {}_1^0e \quad {}_2^3He + {}_2^3He \rightarrow {}_2^4He + 2\,{}_1^1H$$

$$_1^1H + {}_1^2H \rightarrow {}_2^3He \quad\quad {}_2^3He + {}_1^1H \rightarrow {}_2^4He + {}_1^0e$$

One problem with utilizing fusion is the ***4×10^7 Kelvin*** needed to initiate the process.

5.4. Basic Biological Chemistry

Amino Acids and Proteins

If you only remember one thing from this section of Chapter 13, please let it be this:

Proteins are polymers of amino acids.

Amino Acids

An amino acid, surprisingly enough, contains an amine ($-NH_2$) group and an acid ($-COOH$) group.

There are a total of 20 amino acids found in human proteins. Each of these proteins have an amine group, an acid group, and a hydrogen attached to the central, or ***alpha (α)***, carbon. The remaining non- functional, side chain is generally referred to as the ***R-group***.

Think of the **R-group** as the **R**est of the molecule

It is convenient to draw an amino acid as a neutral molecule. However, in aqueous media, they exist as a dipolar ion known as a ***zwitterion*** as shown in Figure 5.7.

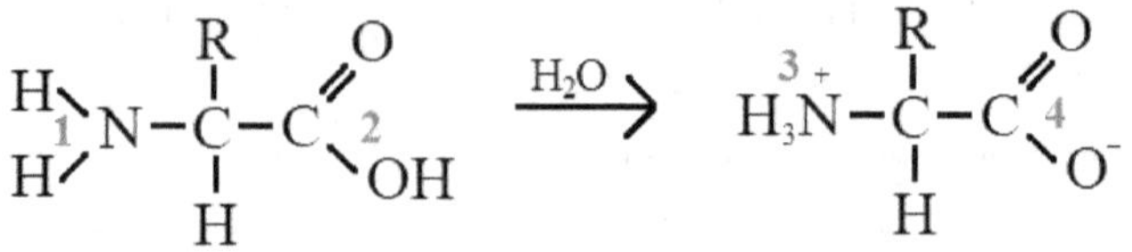

Figure 5.7: The General Structure of Amino Acids

The functional portions of the amino acid in Figure5.7 are numbered in red and are as follows:

1) Amine group
2) Acid group
3) Positivelycharged*ammonium*group
4) Negatively charged *carboxylate* group

The molecule on the right in Figure 5.7 is the *zwitterion* configuration of an amino acid as it exists in aqueous solution.

Remember, there is not an ammonium or carboxylate *ion* present in the zwitterion of an amino acid – no matter how much there appears to be.

Note that the carbon of the amino acid drawn in Figure 5.7 is the *chiral* carbon of the molecule. In Figure 5.7 the R group is defined as a methyl group to represent the amino acid *alanine*.

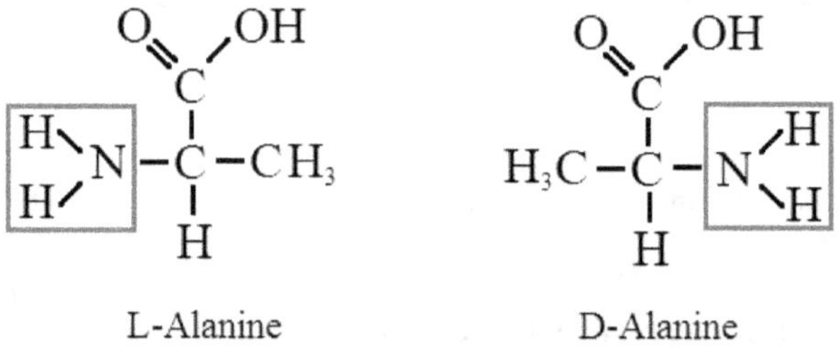

L-Alanine

D-Alanine

Figure 5.8: Stereoisomers of Alanine

Of the twenty amino acids found in proteins all have stereoisomers except the one where the R group is defined as hydrogen. This is the amino acidglycine.

Classifications of Amino Acids

The following tables lists the twenty amino acids found in proteins by their classification, name, symbol, and the structures of their individual R groups. Remember, the drawings in these tables are not amino acids. They are substituent groups for amino acids.

Table 5.2: Nonpolar Amino Acids

Name	Symbol	R Group Structure
Glycine	Gly	$H-$
Alanine	Ala	CH_3-
Valine	Val	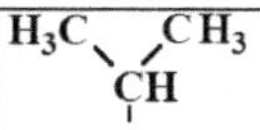 $(H_3C)(CH_3)CH-$
Leucine	Leu	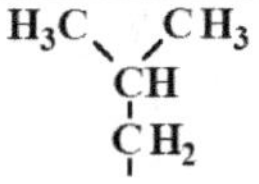 $(H_3C)(CH_3)CH-CH_2-$
Isoleucine	Ile	$H_3C-CH_2CH_2CH_3$
Proline	Pro	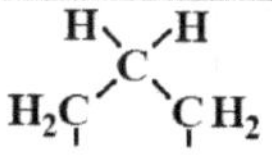 ring: H_2C-CH_2 with $CH(H)(H)$
Methionine	Met	$CH_3-S-CH_2-CH_2-$

Table 5.3: Aromatic Amino Acids

Name	Symbol	R Group Structure
Phenylalanine	Phe	 benzene ring–CH_2-
Tryptophan	Trp	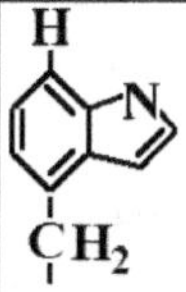 indole ring (H on N)–CH_2-
Tyrosine	Tyr	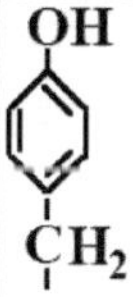 OH–benzene ring–CH_2-

Table 5.4: Polar Amino Acids

Name	Symbol	R Group Structure
Serine	Ser	$\overset{\overset{\textstyle OH}{\mid}}{\underset{\mid}{CH_2}}$
Theronine	Thr	$\overset{\overset{\textstyle CH_3}{\mid}}{\underset{\mid}{H-C=O}}$
Cysteine	Cys	$\overset{\overset{\textstyle SH}{\mid}}{\underset{\mid}{CH_2}}$
Asparagine	Asn	$H_2N-\overset{\overset{\textstyle O}{\|\|}}{\underset{\underset{CH_2}{\mid}}{C}}$
Glutamine	Gln	$H_2N-\overset{\overset{\textstyle O}{\|\|}}{\underset{\underset{\underset{CH_2}{\mid}}{CH_2}}{C}}$

Table 5.5: Acidic Amino Acids

Name	Symbol	R Group Structure
Aspartic Acid	Asp	$\overset{\overset{\textstyle O}{\|\|}}{\underset{\underset{CH_2}{\mid}}{C-OH}}$
Glutamic Acid	Glu	$\overset{\overset{\textstyle O}{\|\|}}{\underset{\underset{\underset{CH_2}{\mid}}{CH_2}}{C-OH}}$

Table 5.6: Basic Amino Acids

Name	Symbol	R Group Structure
Histidine	Hiis	(imidazole ring structure) CH_2
Lysine	Lys	NH_2 — CH_2 — CH_2 — CH_2 — CH_2
Arginine	Arg	NH_2 — $C=NH$ — NH — CH_2 — CH_2 — CH_2

The unattached bond line on the left of the proline R group, attaches to the N of the amine group of the amino acid structure.

Isoelectric Point

The perfectly balanced charged of the zwitterion is not the only charge condition that an amino acid can be in.

In fact, the amino acid is only in the perfectly balanced charge state when the solution the amino acid is in has a certain pH that is known as the **isoelectric point (pI)** for that amino acid.

Table 5.7 shows the three possible charge conditions for an amino acid in relation to solution pH.

Table 5.7: pH vs. pI for Amino Acids

pH vs. pI	Charged Groups Charges	$\Delta[H^+]$	Amino Acid Charge
pH > pI	-COO$^-$, -NH$_2$	$\downarrow$	1-
pH = pI	-COO$^-$, NH$_3$	$\leftrightarrow$	0
pH < pI	-COOH, -NH$_3$	$\downarrow$	1+

Formation of Peptides

The linking of two or more amino acids forms a **peptide**. A peptide has an **amide** bond formed between the -COO$^-$ group of one amino acid to the H_3N^+ group of another amino acid. The H_3N^+ group is called the **N terminal** and the -COO$^-$ group is called the **C terminal**.

$$H_3\overset{+}{N}-CH_2-\overset{\overset{\displaystyle O}{\|}}{C}-O^- + H_3\overset{+}{N}-CH_2-\overset{\overset{\displaystyle O}{\|}}{C}-O^- \longrightarrow H_3\overset{+}{N}-CH_2-\overset{\overset{\displaystyle O}{\|}}{C}-\overset{\overset{\displaystyle H}{|}}{N}-CH_2-\overset{\overset{\displaystyle O}{\|}}{C}-O^- + H_2O$$

Figure 5.8: The Formation of a Peptide

The blue numbered areas in Figure 5.8 are significant as follows:

1) A positively charged amino group and the N terminal

2) A negatively charged carboxylate group and the C terminal

3) The formation of a ***peptide bond***

4) The formation of water in a condensation reaction

Naming Peptides

The naming of a peptide starts from the N terminal amino acid where the -ine, -an, or -ic on the N terminal, and successive amino acids, is replaced with ***-yl*** and the C terminal acid's full name is used.

Problem 5.8

Name the peptide and give its shorthand notation.

$$H_3\overset{+}{N}-CH_2-\overset{\overset{\displaystyle O}{\|}}{C}-\overset{\overset{\displaystyle H}{|}}{N}-\underset{\underset{\displaystyle CH_3}{|}}{CH}-\overset{\overset{\displaystyle O}{\|}}{C}-\overset{\overset{\displaystyle H}{|}}{N}-CH_2-\overset{\overset{\displaystyle O}{\|}}{C}-O^-$$

Glycylalanylglycine (Gly-Ala-Gly)

Problem 5.9

Name the peptide and give its shorthand notation.

$$H_3\overset{+}{N}-\underset{\underset{\displaystyle CH_2OH}{|}}{CH}-\overset{\overset{\displaystyle O}{\|}}{C}-\overset{\overset{\displaystyle H}{|}}{N}-\underset{\underset{\displaystyle CH_3}{|}}{CH}-\overset{\overset{\displaystyle O}{\|}}{C}-\overset{\overset{\displaystyle H}{|}}{N}-\underset{\underset{\displaystyle CH_2OH}{|}}{CH}-\overset{\overset{\displaystyle O}{\|}}{C}-O^-$$

Serylalanylserine (Ser-Ala-Ser)

Problem 5.10

Name the peptide and give it's shorthand notation.

$$H_3\overset{+}{N}-CH_2-\overset{\overset{\displaystyle O}{\|}}{C}-\overset{\overset{\displaystyle H}{|}}{N}-CH_2-\overset{\overset{\displaystyle O}{\|}}{C}-\overset{\overset{\displaystyle H}{|}}{N}-\underset{\underset{\displaystyle CH_3}{|}}{CH}-\overset{\overset{\displaystyle O}{\|}}{C}-\overset{\overset{\displaystyle H}{|}}{N}-\underset{\underset{\displaystyle CH_2OH}{|}}{CH}-\overset{\overset{\displaystyle O}{\|}}{C}-O^-$$

Glycylglycylalanylserine (Gly-Gly-Ala-Ser)

In the drawings of peptides above it is easy to see the peptide bonds stick up. If you count n peptide bonds then the peptide is constructed of $n + 1$ amino acids. Therefore the name should have $n + 1$ parts.

Levels of Protein Structure

The **primary structure** of a protein is the type of structures we have seen so far. Simply line structures with elemental symbols and bonds. The drawing below is the primary structure of arginylhistidinyllysine (Arg-His-Lys).

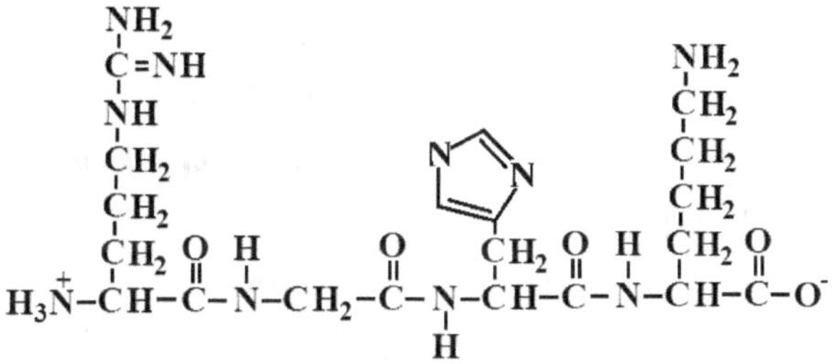

Figure 5.8: A Primary Protein Structure

The **secondary structure** of a protein is that structure formed when two proteins are linked to one another through **hydrogen bonding.**

The best example of this intermolecular bonding is the bonds between the two helices in the double helix structure of DNA.

Figure 5.9 shows a representation of a strand of DNA. Hydrogen bonding between a hydrogen on an amine group and an oxygen on a carboxylate group is what holds the two helices of DNA together.

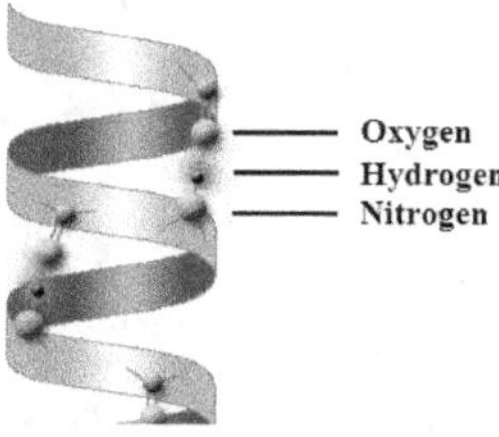

Figure 5.9: A Secondary Protein Structure

The need for a secondary structure is that there is a way to view the structure of a molecule. The helical shape of DNA is evident in Figure 5.9.

Figure 5.10 shows another secondary structure evident in DNA as the ***beta-pleated sheet.***

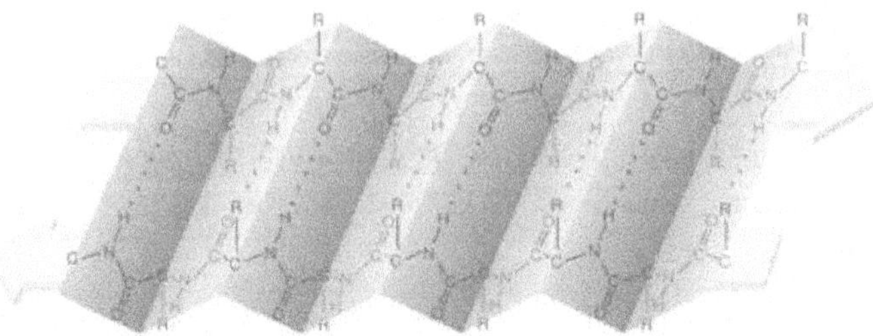

Figure 5.10: The Beta–Pleated Sheet

The very long peptide chains of proteins are capable of ***intramolecular interactions***, of which hydrogen bonding is only one example.

These interactions along the peptide chain result in a certain three dimensional shape for the protein. Figure 5.11 shows a tertiary form of a protein with four distinct intramolecular interactions.

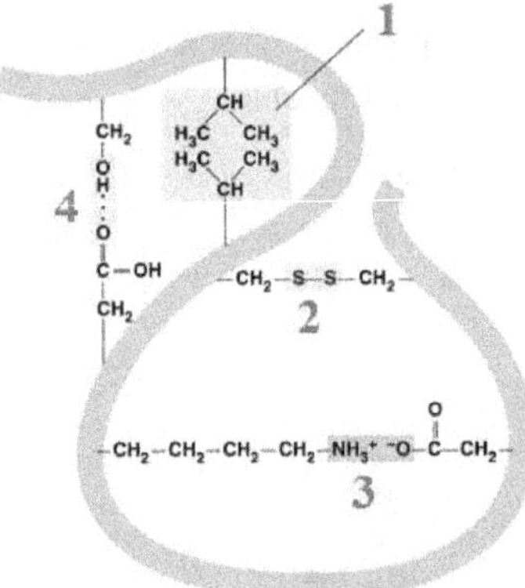

Figure 5.11: A Tertiary Protein Structure

The distinctions of the four intramolecular interactions are as follows:

1) A hydrophobic (water fearing) reaction resulting in a van der Waals attraction

2) A disulfide bond

3) An ionicbond

4) A hydrogen bond

When a biologically active protein consists of two or more polypeptide subunits, the structural level is referred to as a **_quaternary structure_**.

Hemoglobin is a globular protein consisting of four polypeptide chains, two α-chains and two β-chains. The subunits are held together by the same interactions that stabilize tertiary protein structures.

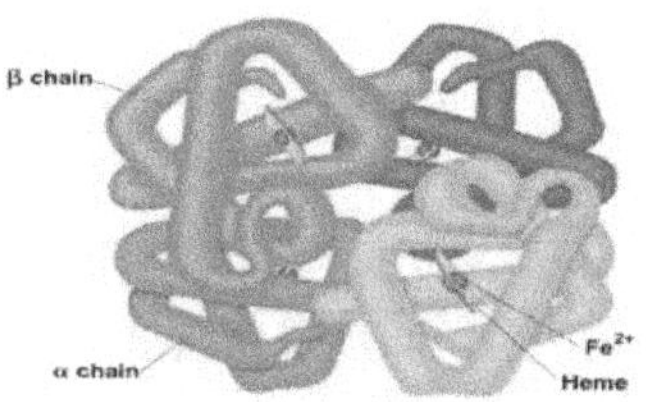

Figure 5.12: A Quaternary Protein Structure

Hemoglobin is known as a **_globular protein_**. Globular proteins are roughly spherical in shape because the peptide chains fold over each other.

Denaturation of Proteins

Denaturation of proteins occurs when there are disruptions of any of the bonds that establish the secondary, tertiary, or quaternary structures of a protein.

The following list the effects of certain conditions and chemicals on the internal bonds of proteins.

Excessive **_heat_**, in excess of 50°C, can disrupt hydrogen bonding as well as the hydrophobic attraction between nonpolar side chains.

Acids and **_bases_** disrupt hydrogen bonding in polar side chains. They can also disrupt the bonding in salt bridges.

Organic compounds can disrupt the hydrophobic interactions between nonpolar side chains.

Heavy metals, such as silver and lead, can disrupt disulfide bonds through the formation of ionic bonds with the sulfur.

Enzymes

Enzymes are the biological catalysts needed for most of the chemical reactions taking place in the body. In general they act to lower the activation energy for a particular reaction. Enzymes are named by replacing the end of the reaction or reacting compound with **_-ase_**.

Figure 5.13 shows the lower activation energy for a reaction made possible by the presence of an enzyme catalyst.

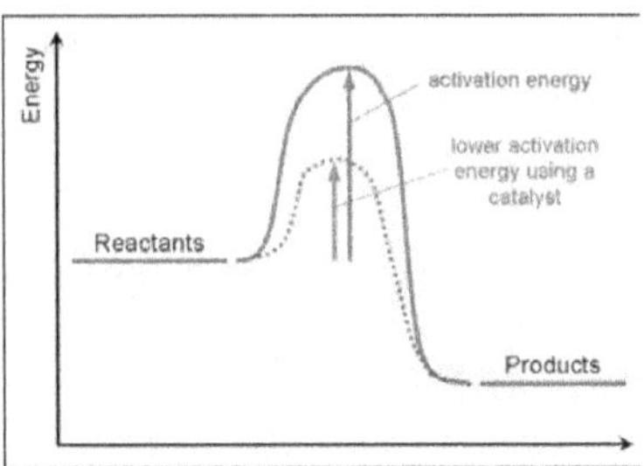

Figure 5.13: Enzyme Catalyst

The following is a representation of the basic enzyme catalyzed reaction. An enzyme (**E**) will form with a substrate (**S**) to form an enzyme substrate complex (**ES**). The enzyme substrate (ES) complex will react with reactant (**R**) to form a product (**P**) and the enzyme.

$$E + S \leftarrow ES$$
$$ES + R \leftarrow P + E$$

Enzyme Action

Nearly all enzymes are globular proteins. Each has a unique shape that recognizes and binds a small group of molecules, called **substrates**.

In this way, the tertiary structure of enzymes play an important role in catalysis reactions. Figure 5.14 is a graphic representation of an enzyme catalyzed reaction.

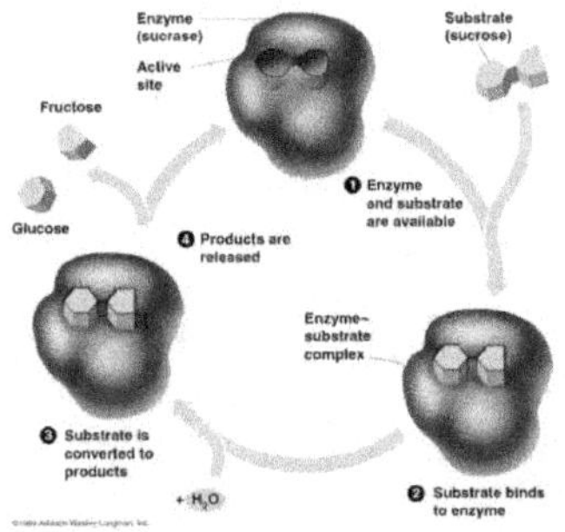

Figure 5.14: An Enzyme Catalyzed Reaction

Here **sucrose** is catalytically separated into **fructose** and **glucose** through a **sucrase** catalyzed reaction.

The Induced Fit Model

The positioning of the sucrose molecule onto the sucrose enzyme is what is known as the **lock and key** model; where the sucrose molecule is the key and sucrose enzyme is the lock.

In the theory of the induced fit model it is believed that an active site in the enzyme is able to adjust to better accommodate the reaction.

Figure 5.15 is a graphic demonstration of the induced fit model.

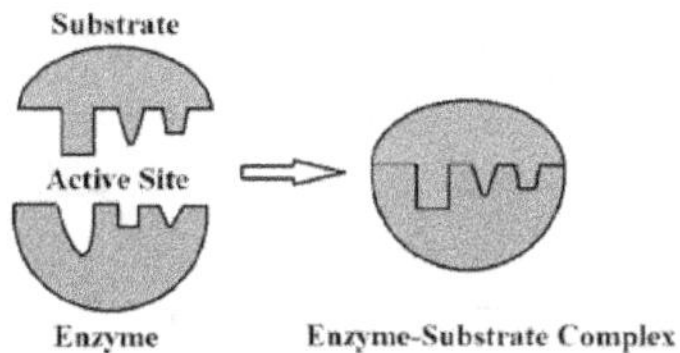

Figure 5.15: The Induced Fit Model

Enzyme Inhibitors

Enzyme inhibitors cause enzymes to lose catalytic activity. The two kinds of enzyme inhibitors are the **competitive inhibitor** and the **noncompetitive inhibitor**.

In the Figure 5.16 below, a competitive inhibitor is preventing the binding of a substrate.

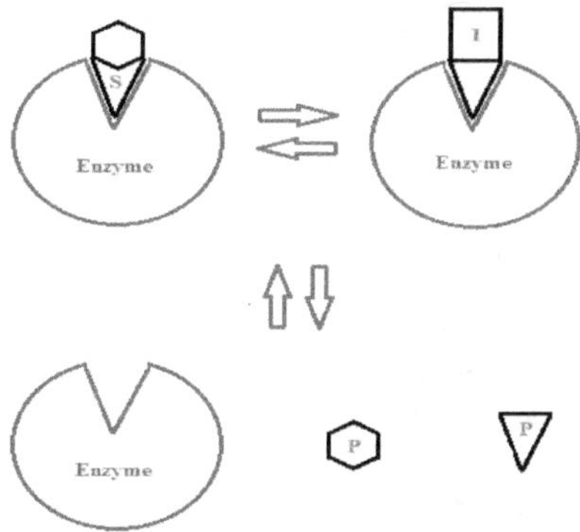

Figure 5.16: A Competitive Enzyme Inhibitor

Factors Affecting Enzyme Activity

It should be evident that among the factors affecting the activity of an enzyme are the **concentration of the enzyme**, the **concentration of the substrate**, and the presence of any **enzyme inhibitors**.

Two remaining factors that affect the activity of enzymes are **pH** and **temperature.** Most enzymes are at optimal activity at 37°C or human body temperature. The pH for the optimal activities of different enzymes varies widely.

Table 5.8 lists the optimal activity pH for some common enzymes.

Table 5.8: pH for Optimal Enzyme Activity

Enzyme	Optimal pH
Pepsin	1.5–1.6
Lipase (stomach)	4.0–5.0
Invertase	4.5
Amylase (malt)	4.6–5.2
Lipase (castor oil)	4.7
Maltase	6.1–6.8
Amylase (pancreas)	6.7–7.0
Catylase	7.0
Urease	7.0
Trypsin	7.8–8.7
Lipase (pancreas)	8.0

Enzyme Cofactors

A simple enzyme consists of only proteins. However, many enzymes require small molecules or metal ions, known as **cofactors**, to properly catalyze reactions. The table below lists a series of metal ions used as cofactors. When organic molecules are used as cofactors they are known as **coenzymes**. Table 5.9 lists some metal ions, and vitamins with their enzymes and their functions.

Table 5.9: Metal Ions and Vitamins Enzyme Cofactors

Cofactor	Coenzyme/Enzyme	Function
niacin	nicotinamide adenine dinucleotide (NAD^+)	oxidation or hydrogen transfer
riboflavin	flavin adenine dinucleotide (FAD)	oxidation or hydrogen transfer
pantothenic acid	coenzyme A (CoA)	acetyl group carrier
vitamin B-12	coenzyme B-12	transfers methyl groups
vitamin B-1 (thiamin)	thiaminpyrophosphate (TPP)	transfers aldehyde groups
Cu^{2+}	cytochrome oxidase	oxidation-reduction
Fe^{2+}	catalase	oxidation-reduction
Fe^{3+}	cytochrome oxidase	oxidation-reduction
Mg^{2+}	glucose-6 phosphate	phosphate ester hydrolyzation
Mn^{2+}	arginase	electron removal
Ni^{2+}	Urease	amide hydrolyzation

Nucleic Acids: DNA and RNA

Components of Nucleic Acids

Two closely related types of nucleic acids, *deoxyribonucleic acid (DNA)* and *ribonucleic acid (RNA)* are both unbranched polymers of repeating monomer units known as *nucleotides*.

DNA molecules contain up to several million nucleotides and RNA molecules contain several thousand nucleotides. These nucleotides are all made up of a *base*, a *five-carbon sugar*, and a *phosphate group*.

The Five Bases of DNA and RNA

The five bases found in nucleic acids are of two types: pyrimidines and purines. The three pyrimidines used in the construction of nucleic acids are *cytosine*, *thymine*, and *uracil*.

The two purines used in the construction of nucleic acids are *adenine* and *guanine*.

Figure 5.17 shows the structures of the five bases used to construct the nucleic acids in DNA and RNA. It is essential to remember that *thymine is unique to DNA* and *uracil is unique to RNA*.

That is to say, you will never see an RNA molecule containing thymine. Likewise, you will never see a DNA molecule containing uracil.

Pyrimidines

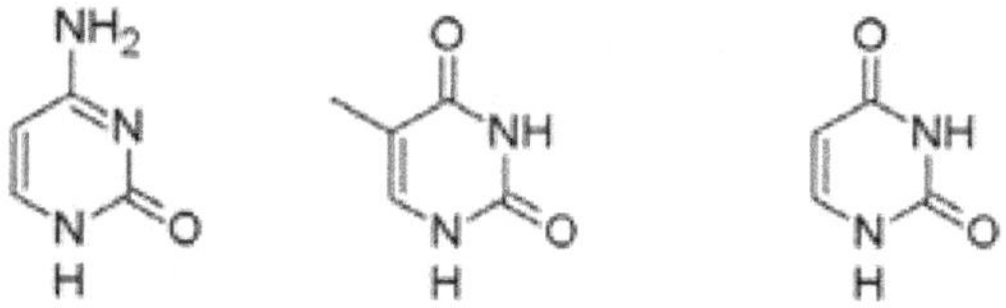

Cytosine   Thymine   Uracil

Purines

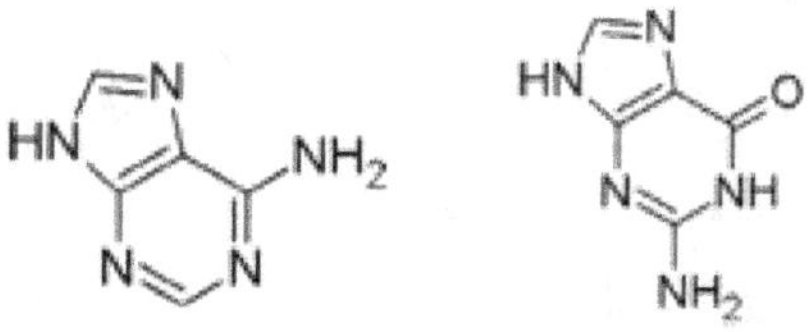

Adenine Guanine

Figure 5.17: The Fives Bases of DNA and RNA

The Two Sugars of DNA and RNA

The two sugars used in the construction of DNA and RNA are deoxyribose and ribose.

Figure 5.18 shows the structure of ribose followed by the structure of deoxyribose where there is clearly a hydroxide group missing in comparison to the ribose structure.

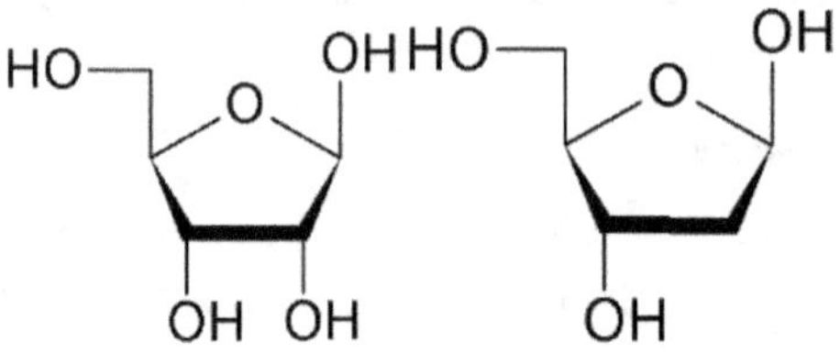

Ribose Deoxyribose

Figure 5.18: The Two Sugars of DNA and RNA

Nucleosides and Nucleotides

A **nucleoside** is the term used when a base forms a bond with the C1 carbon in the sugar.

A **nucleotide** is the term used when the nucleoside bonds a phosphate group to the C5 -OH group on the sugar.

The example of each are shown in Figure 5.19.

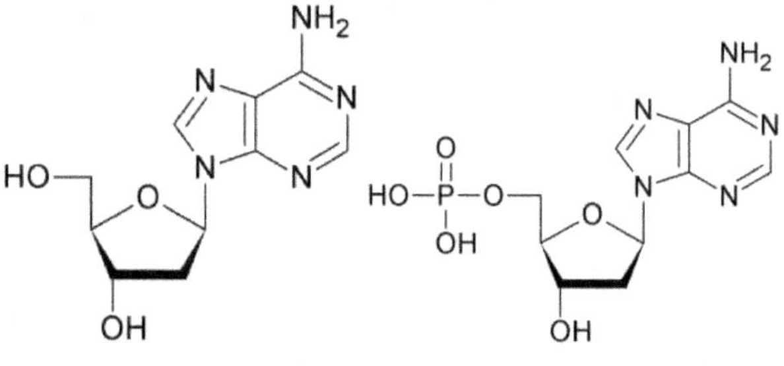

A Deoxyribonucleoside   A Deoxyribonucleotide

Figure 5.19: A Nucleoside and A Nucleotide

Naming Nucleosides and Nucleotides

Nucleosides and nucleotides containing purine end with *–osine*. Nucleosides and nucleotides containing pyrimidine end with *–ine*. **Deoxy** is added to the front of all DNA nucleosides and nucleotides. Table 5.10 lists the four nucleosides and the four nucleotides of DNA and RNA.

Table 5.10: Nucleosides and Nucleotides of DNA and RNA

The Bases	The Nucleosides	The Nucleotides
Deoxyribonucleic Acids (DNA)		
Adenine (A)	Deoxyadenosine (A)	Deoxyadenosine 5'-monophosphate (dAMP)
Guanine (G)	Deoxyguanosine (G)	Deoxyguanosine 5'-monophosphate (dGMP)
Cytosine (C)	Deoxycytidine (C)	Deoxycytidine 5'-monophosphate (dCMP)
Thymine (T)	Deoxythymidine (D)	Deoxythymidine 5'-monophosphate (dUMP)
Ribonucleic Acids (RNA)		
Adenine (A)	Adenosine (A)	Adenosine 5'-monophosphate (AMP)
Guanine (G)	Guanosine (G)	Guanosine 5'-monophosphate (GMP)
Cytosine (C)	Cytidine (C)	Cytidine 5'-monophosphate (CMP)
Uracil (U)	Uridine (U)	Uridine 5'-monophosphate (UMP)

Primary Structure of Nucleic Acids

The nucleic acids consist of polymers of many nucleotides in which the 3'-OH group in the sugar of one nucleotide is bonded, through the phosphate group, to the 5'-carbon atom of the sugar of another nucleotide in what is known as the ***phosphodiester bond***.

Figure 5.20 shows a phosphodiester bond between adenosine 5'-monophosphate (AMP) and guanosine 5'-monophosphate (GMP).

Figure 5.20: Phosphodiester Bond between AMP and GMP

The DNA Double Helix

The double helix of the DNA molecule is formed like a spiral staircase. The sugar-phosphate backbones form the railing while the steps are formed by hydrogen bonding of the bases in the two helices.

These hydrogen bonded bases are known as ***complimentary base pairs.***

The base pairs formed are exclusive to ***adenine to thymine*** by two hydrogen bonds and ***guanine to cytosine*** by three hydrogen bonds.

Figure 5.21 shows the double helical shape of DNA

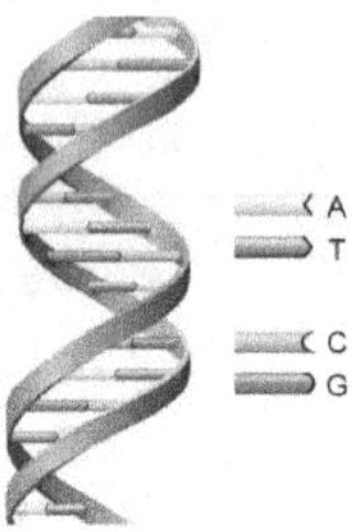

Figure 5.21: The DNA Double Helix

Please note the restriction to hydrogen bonding between adenine and thymine and between cytosine and guanine.

Let us take a closer look at hydrogen bonding that occurs between the complementary base pairs in the DNA molecule.

Figure 5.22 shows an expansion of the "ladder like" construction of the DNA molecule in between the two helices.

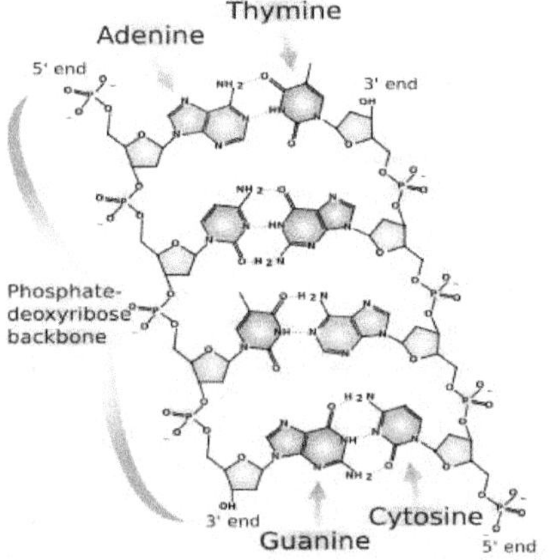

Figure 5.22: Hydrogen Bonding in DNA Base Pairs

DNA Replication

In DNA replication, the separate strands of the parent DNA are the templates for the synthesis of complimentary strands, which produce two exact copies of the parent DNA.

Figure 5.23 shows the replication of a DNA strand.

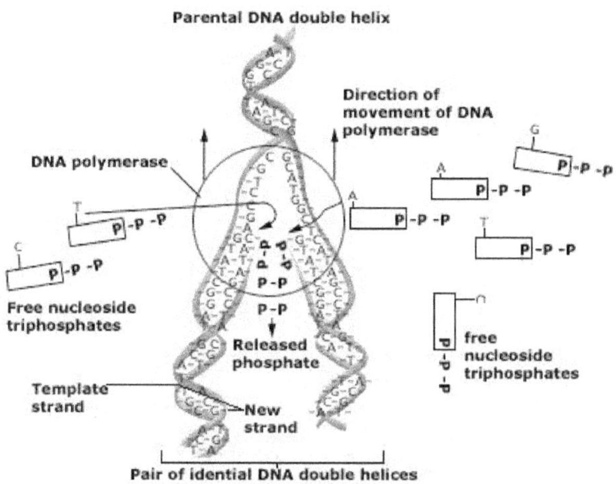

Figure 5.23: DNA Replication

Protein Synthesis

RNA and the Genetic Code

There are three major types of RNA in the cells of the body.

Ribosomal RNA (rRNA): rRNA serves as the sites for the synthesis of proteins.

Messenger RNA (mRNA): mRNA carries protein synthesis information from the DNA to the ribosomes in the cell cytoplasm.

Transfer RNA (tRNA): tRNA interprets genetic information in mRNA and provides amino acid specific information for protein synthesis.

RNA and Protein Synthesis

Transcription is the process where information from a gene in the DNA is copied to mRNA for the synthesis of a protein.

Translation is the process where tRNA converts the information in the mRNA into the proper sequence of amino acids needed for protein synthesis.

Figure 5.24 shows the progression from DNA through messenger RNA to the synthesis of a protein.

Figure 5.24: Protein Synthesis

The Genetic Code

The genetic code consists of series of three nucleotides (triplets) in the mRNA known as **codons**.

An example codification is a triplet of uracil (UUU) which produces a peptide that contains only phenylalanine.

A total of 64 codons are possible from the combinations of adenine (A), guanine (G), cytosine (C), and uracil (U).

Table 5.12: mRNA Codons and Resultant Amino Acids

1st * 3rd	Second Letters (Symbol)			
	U	C	A	G
U * U	UUU (Phe)	UCU (Ser)	UAU (Tyr)	UGU
U * C	UUC (Phe)	UCC (Ser)	UAC (Tyr)	UGC
U * A	UUA (Leu)	UCA (Ser)	UAA	UGA
U * G	UUG (Leu)	UCG (Ser)	UAG	UGG
C * U	CUU (Leu)	CCU (Pro)	CAU (His)	CGU
C * C	CUC (Leu)	CCC (Pro)	CAC (His)	CGC
C * A	CUA (Leu)	CCA (Pro)	CAA	CGA
C * G	CUG (Leu)	CCG (Pro)	CAG (Gln)	CGG
A * U	AUU (Ile)	ACU (Thr)	AAU	AGU (Ser)
A * C	AUC (Ile)	ACC (Thr)	AAC	AGC (Ser)
A * A	AUA (Ile)	ACA (Thr)	AAA	AGA
A * G	AUG (Start)	ACG (Thr)	AAG	AGG
G * U	GUU (Val)	GCU (Ala)	GAU	GGU (Gly)
G * C	GUC (Val)	GCC (Ala)	GAC	GGC (Gly)
G * A	GUA (Val)	GCA (Ala)	GAA (Glu)	GGA (Gly)
G * G	GUG (Val)	GCG (Ala)	GAG (Glu)	GGG (Gly)

The Start and Stop results for certain of the codons in Table 5.12 are for starting and stopping protein sequences.

Translation of mRNA by tRNA

Each tRNA contains a loop called an **anticodon**, which is a triplet of bases that complements a codon in an mRNA.

tRNA is activated for protein synthesis by the tRNA synthetase enzyme. The synthetase uses the anticodons to attach the correct amino acid to the tRNA stem.

Figure 5.25 shows the anticodon loop of tRNA.

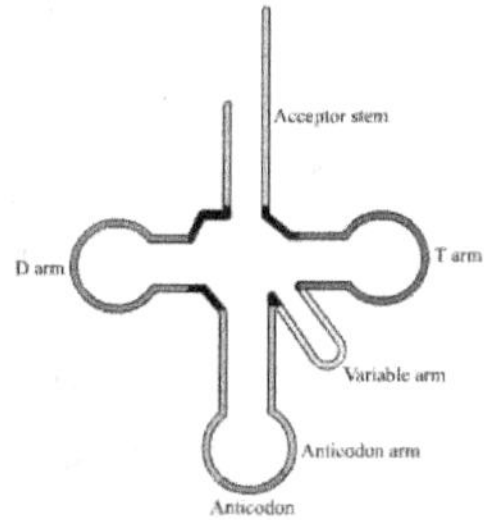

Figure 5.25: tRNA Anticodon Loop

Genetic Mutations

A **mutation** is a change in the DNA nucleotide sequence that alters the sequence of amino acids, which may alter the structure and function of a protein.

A **substitution** mutation occurs when one base in the coding strand of DNA is replaced with another. A **frame shift** mutation occurs when a base is added or deleted from the coding strand of the DNA. Figure 5.26 shows a graphic demonstration of substitution and both addition and deletion mutations.

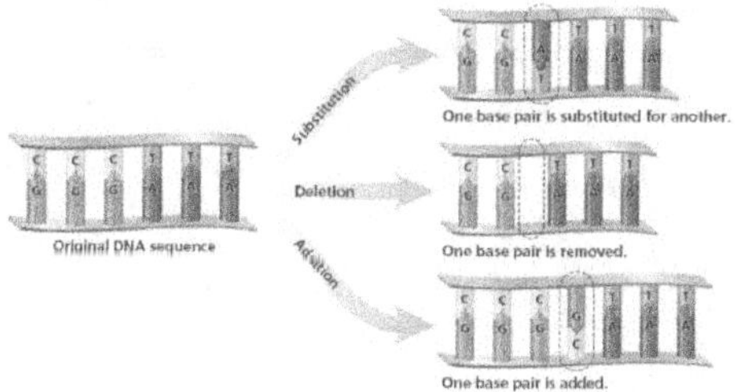

Figure 5.26: Genetic Mutations

Metabolism and Energy Production

Types of Metabolism

The term ***metabolism*** refers to all the chemical reactions that provide the energy and materials necessary for cell growth. There are two types of metabolic reactions.

In ***catabolic reactions*** complex molecules are broken down into smaller ones resulting in a release of energy.

Anabolic reactions utilize available energy in cells to produce larger molecules from smaller ones.

Stages of Metabolism

There are three stages of metabolism.

In Stage 1, the processes of ***digestion*** break down larger macromolecules into smaller units.

In Stage 2, these molecules are further broken down to 2 and 3 carbon species, such as pyruvate and acetyl coenzyme A (CoA), through the processes of ***oxidation*** and ***hydrolysis***.

In Stage 3, the ***oxidation*** of materials to CO_2 and H_2O provides energy for adenosine triphosphate (ATP) synthesis.

Figure 5.27 shows all of the intricacies of metabolism. Most of these will be covered in the ensuing pages.

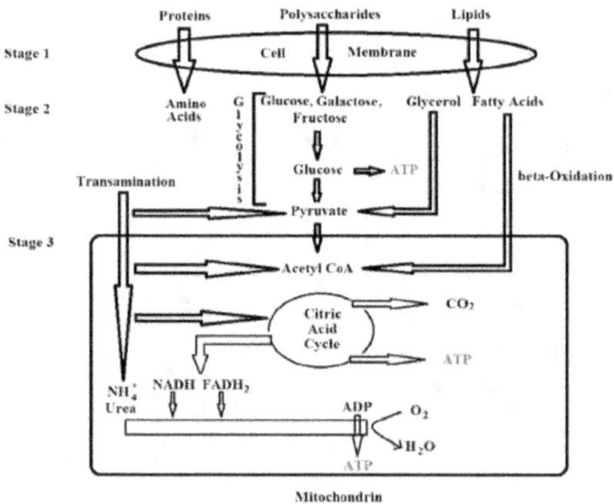

Figure 5.27: Stages of Metabolism

ATP and Energy

When the food we eat is metabolized, the energy released from the oxidation process is stored in the **adenosine triphosphate (ATP)** molecule. When ATP is hydrolyzed the products of the reaction are adenosine diphosphate (ADP), a phosphate group, and **7.3 kcal** per mole of ATP.

Figure 5.28 shows the molecular structure of adenosine triphosphate (ATP).

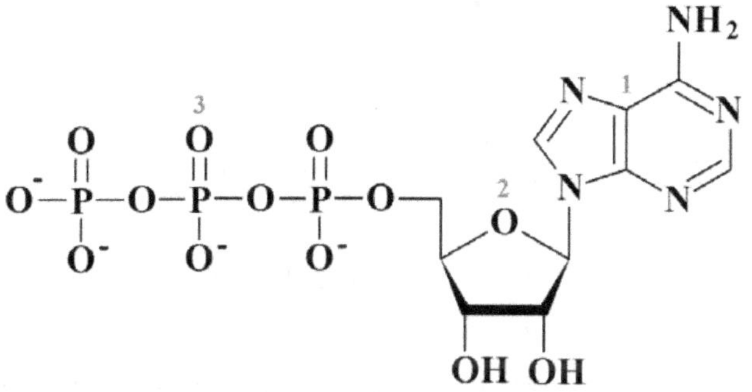

Figure 5.28: Adenosine Triphosphate (ATP)

As a point of review, let's interpret the red numbers from Figure 5.28.

1. The adenine molecule
2. The ribose molecule (Together 1 and 2 comprise the adenosine molecule.)
3. The ionic triphosphate chain

All together these components make adenosine triphosphate.

Digestion of Carbohydrates

The salivary enzyme **amylase** hydrolyzes α-glycosidic bonds in amylose and amylpectin to produce maltose, glucose, and the smaller polysaccharides dextrins.

In the small intestines, at a pH of around 8, an α-amylase (produced in the pancreas) hydrolyzes the remaining polysaccharides to maltose and glucose.

Enzymes produced in the mucosal cells of the small intestines can hydrolyze maltose, glucose, and sucrose.

Figure 5.29 shows the digestion of carbohydrates from ingestion to the deposit of glucose into the bloodstream.

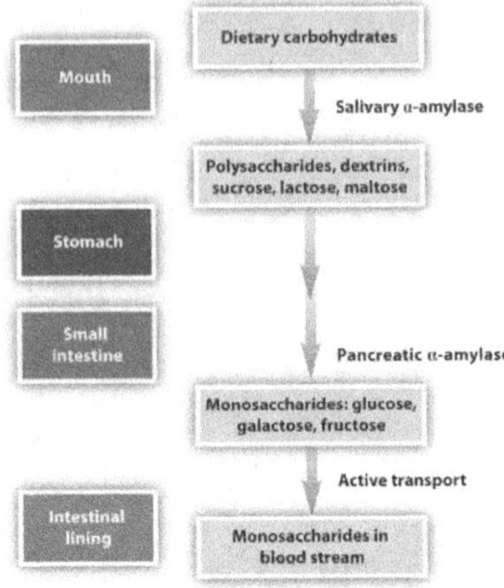

Figure 5.29: Digestion of Carbohydrates

Maintaining Blood Glucose Levels

After eating, the blood *glucose* level is elevated. The pancreas secretes the hormone *insulin* which increases the flow of glucose into muscle and adipose tissue. There the insulin converts the glucose to *glycogen* which is stored in the liver.

When you are fasting throughout the night, blood glucose levels drop. This stimulates the pancreas to secrete another hormone *glucagon*, which stimulates the breakdown of glycogen in the liver in order to secrete glucose.

Digestion of Triacylglycerols

In the small intestines, a process known as *emulsification* uses bile salts from the gall bladder to form fat micelles. Then *lipases* produced in the pancreas hydrolyze the micelle fat to form free fatty acids and amonoacylglycerol.

These products are reformed in the intestinal lining to triacylglycerols coated with proteins to form lipoproteins known as *chylomicrons*. In the cells, enzymes hydrolyze the triacylglycerides to glycerol and free fatty acids.

Digestion of Proteins

In the stomach, hydrochloric acid (HCl), at pH=2, denatures proteins and activates enzymes, such as *pepsin*, that begin to hydrolyze peptide bonds. The polypeptides move from the stomach into the small intestines where the enzymes *trypsin* and *chymotrypsin* complete the hydrolysis of the peptides into amino acids. These amino acids are then absorbed through the intestinal walls into the bloodstream for transport to the cells.

Metabolic Pathway Coenzymes

Before we look at some important coenzymes, let's review the concepts of ***oxidation*** and ***reduction***.

Oxidation is a loss of H and electrons. When an enzyme catalyzes oxidation hydrogen atoms and electrons are removed from a substrate.

Reduction is a gain of H and electrons. When hydrogen atoms and electrons are picked up by a coenzyme, it is reduced.

Nicotinamide adenine dinucleotide ***(NAD^+)*** is an important coenzyme formed by the nicotinamide group from the vitamin niacin, which is bonded to adenosine phosphate.

NAD+ participates in producing C=O bonds through the oxidation of alcohols to aldehydes and ketones. Note that Figure 5.30 shows the conversion of NAD^+ to NADH is a reduction for the coenzyme.

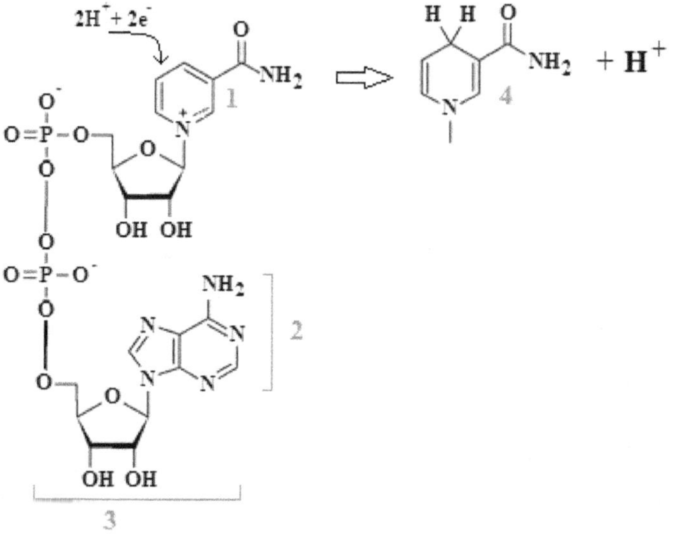

$$NAD^+ + 2H^+ + 2e^- \implies NADH + H^+$$

Figure 5.30: NAD+ Conversion to NADH

Let's interpret the red numbers from Figure 5.30

1. The nicotinamide from the niacin molecule
2. Adenosine diphosphate (ADP)
3. The NAD^+ ion
4. The NADH molecule

Flavin adenine dinucleotide **(FAD)** is a coenzyme containing riboflavin (vitamin B_2) and adenosine diphosphate (ADP).

Riboflavin consists of a sugar alcohol (ribitol) and flavin. Two nitrogen atoms in the FAD receive hydrogens and are reduced to $FADH_2$.

Figure 5.31 shows the conversion of FAD to $FADH_2$.

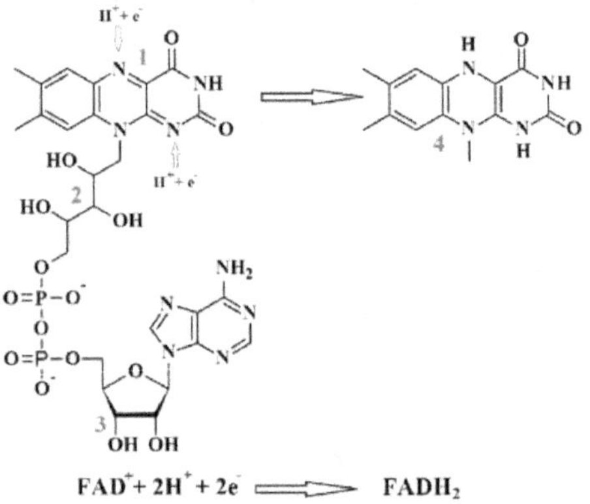

$$FAD^+ + 2H^+ + 2e^- \longrightarrow FADH_2$$

Figure 5.31: FAD Conversion to $FADH_2$

Let's interpret the red numbers from Figure 5.31

1. The flavin molecule
2. The ribitol molecule
3. The ADP ion
4. The $FADH_2$ molecule

Coenzyme A (CoA) is made up of pantothenic acid (vitamin B_5) adenosine diphosphate (ADP) and aminoethanethiol. CoA prepares small groups like the acetyl group for enzyme action.

Figure 5.32 shows the structure of Coenzyme A (CoA).

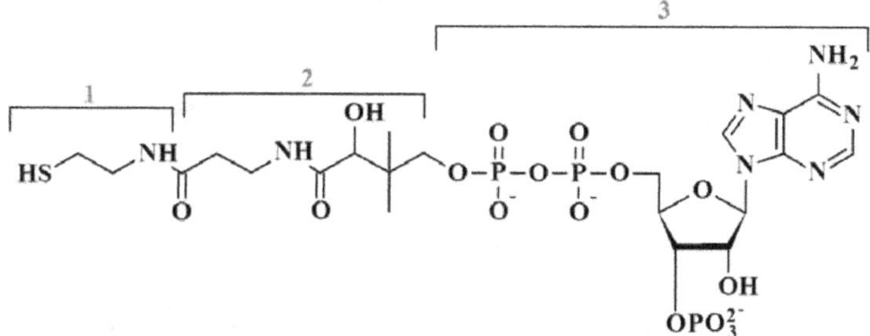

Figure 5.32: Coenzyme A (CoA)

Let's interpret the red numbers from Figure 5.32

1. The molecule aminoethanethiol
2. The molecule pantothenic acid
3. The ion phosphorylated ADP

Glycolysis, an anaerobic process (no oxygen required), is a Stage 2 process where a six carbon glucose is broken down into two three carbon molecules such as pyruvate.

Figure 5.33 shows the overall process of the conversion of glucose to NADH in three distinct stages.

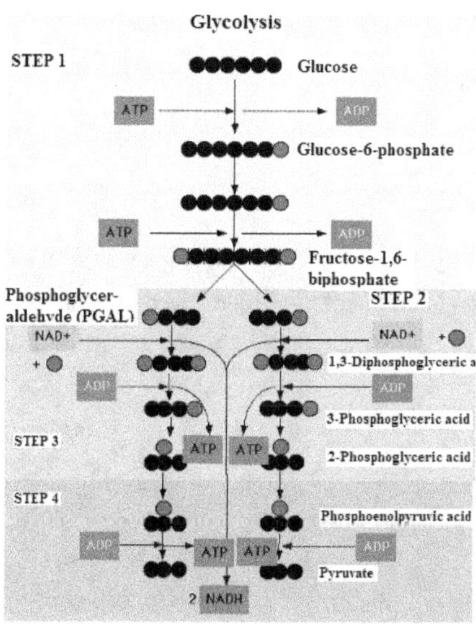

Figure 5.33: Stages of Glycolysis

The Flavin Mononucleotide (FMN) Coenzyme System

FMN (flavin mononucleotide) is a coenzyme derived from riboflavin (vitamin B_2). In riboflavin, the flavin is connected to ribotol (the sugar alcohol of ribose). Here FMN is reduced to $FMNH_2$.

Figure 5.34 shows the reduction of FMN to $FMNH_2$.

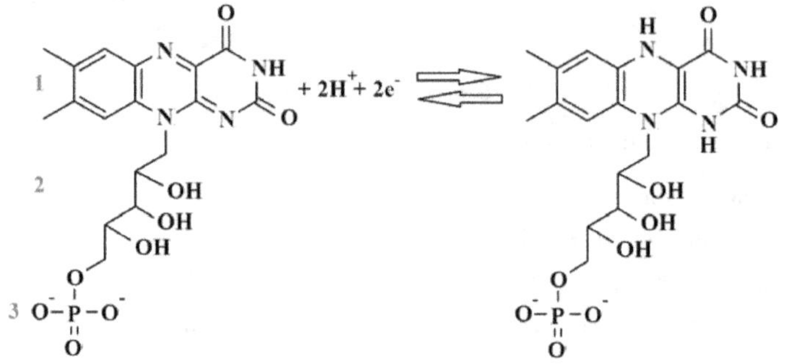

FMN FMNH$_2$

Figure 5.34: A Flavin Mononucleotide Reduction

Some constituent identifications for Figure 5.34 follow.

1. The flavinmolecule
2. The ribitol molecule
3. The riboflavin phosphate ion

The Coenzyme Q (CoQ) System

Coenzyme Q (Q or CoQ) is derived from quinone as seen below. When the two ketone groups accept hydrogens and electrons they are reduced to alcohols.

Figure 5.35 shows the reduction of CoQ.

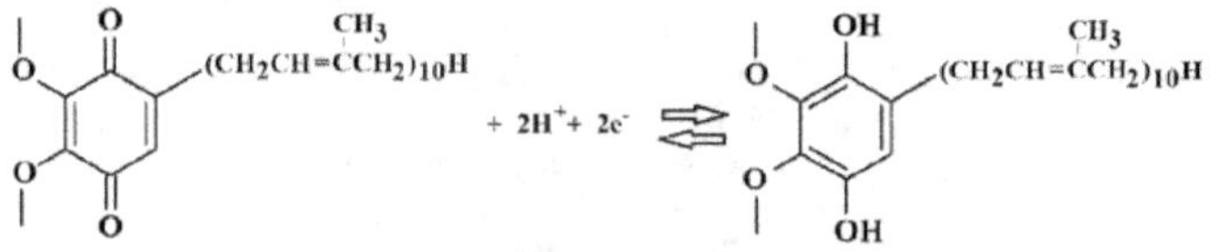

CoQ CoQ - Reduced

Figure 5.35: A Coenzyme Q Reduction

The Cytochrome (cyt) System

Cytochromes (cyt) are proteins that carry an iron in a heme group. In each of the cytochromes (cyt b, cyt c_1, cyt c, cyt a, and cyt a_3) Fe^{2+} is oxidized back to Fe^{3+} through the loss of an electron.

Figure 5.36 shows the reduction of cytochrome c (cyt c).

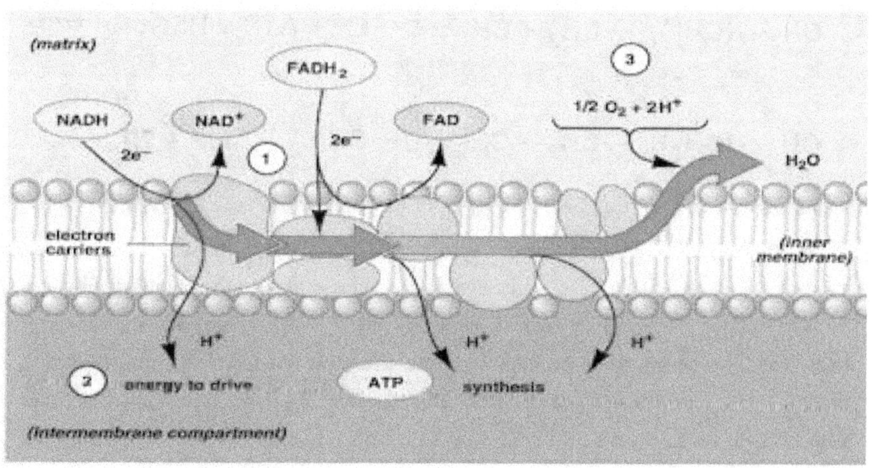

Figure 5.36: A Cytochrome c Reduction

In the inner membrane of the mitochondria, **CoQ** and **cyt c** are not firmly attached to the membrane. As such they are considered as mobile carriers that transfer electrons between protein complexes.

Figure 5.37 shows the direction of electron transfer in the inner mitochondrial membrane.

Figure 5.37: Inner Mitochondrial Electron Transfer

Oxidative Phosphorylation and ATP

Oxidative phosphorylation is the process of using the energy of the electrons from the oxidation of substrates to produce ATP in the cell. The process occurs in the intermitochondrial membrane of the cell.

Oxidation of Fatty Acids

A large amount of energy is obtained by the oxidation of fatty acids in the mitochondria. These fatty acids undergo **β-oxidation** and yield acetyl CoA. This process continues, removing two carbons at a time to make more acetyl CoA, until the fatty acid is completely degraded. The acetyl CoA that is produced can then enter the citric acid cycle

That being said, the numbering of atoms in the molecule starts with the carbonyl carbon as number 1. Rather than referring to the carbon just to the left of the carbonyl carbon as one away from the carbonyl carbon it is referred to as **alpha (α) to** the carbonyl carbon. The next carbon to the left is then referred to as **beta (β)** to the carbonyl carbon.

When **fatty acid activation** occurs, in the cytosol of the cell, an acyl group is added to the fatty acid. This activation process combines a fatty acid and CoA to yield fatty acyl CoA. The energy for this process comes from hydrolysis of ATP to produce adenosine monophosphate (AMP) and two inorganic phosphorous atoms.

$$CH_3-(CH_2)_{14}-CH_2-CH_2-\overset{\overset{\textstyle O}{\|}}{C}-O^- + ATP + HSCoA \implies$$

$$CH_3-(CH_2)_{14}-CH_2-CH_2-\overset{\overset{\textstyle O}{\|}}{C}-SCoA + AMP + 2P$$

Urea Cycle

The final product of amino acid degradation is NH_4^+. NH_4^+ is toxic if it is allowed to accumulate in the body. Through a series of reactions if it is allowed to accumulate in the body. Through a series of reactions known as the **urea cycle**, the NH_4^+ is detoxified and converted to urea which is removed from the body by urine. If urea is not removed from your body, as measured by blood urea nitrogen (BUN) procedures as extreme as hemodialysis may be required.

$$2NH_4^+ + CO_2 \quad H_2NCNH_2 + 2H^+ + H_2O$$

Questions

Part-A

1. DEFINE energy
2. What is kinetic energy?
3. What is potential energy?
4. Define exothermic and endothermic reactions
5. What is anode and cathode?
6. What is Electrolysis?
7. What is Electroplating
8. Define Corrosion
9. What is Nuclear Reactions
10. What are the Distinctions of Nuclear Reactions?
11. Define Half-Life
12. Define Fission
13. Define Fusion
14. Define Amino Acids

Part-B

1. Explain in detail the concept of heat and energy?
2. Briefly explain Hess law?
3. Short note on galvanic cell?
4. Briefly explain Types of Radiation?